# All About Vegetables

Created and designed by
the editorial staff of
ORTHO BOOKS

Writer
**Walter L. Doty**

Revision Editor
**Anne Reilly**

Illustrator
**Ron Hildebrand**

Designer
**Gary Hespenheide**

## Ortho Books

**Publisher**
Edward A. Evans

**Editorial Director**
Christine Jordan

**Production Director**
Ernie S. Tasaki

**Managing Editors**
Michael D. Smith
Sally W. Smith

**System Manager**
Linda M. Bouchard

**National Sales Manager**
J. D. Gillis

**National Accounts Manager—Book Trade**
Paul D. Wiedemann

**Marketing Specialist**
Dennis M. Castle

**Distribution Specialist**
Barbara F. Steadham

**Operations Assistant**
Georgiann Wright

**Administrative Assistant**
Francine Lorentz-Olson

**Technical Consultant**
J. A. Crozier, Jr., Ph.D.





Address all inquiries to:
Ortho Books
**Box 5006**
**San Ramon, CA 94583-0906**





8 9

94 95

ISBN 0-89721-222-3
Library of Congress Catalog Card Number 90-80068

THE SOLARIS GROUP
**2527 Camino Ramon**
**San Ramon, CA 94583**

## Acknowledgments

**Photo Editor**
Sarah Bendersky

**Copy Chief**
Melinda E. Levine

**Editorial Coordinator**
Cass Dempsey

**Copyeditor**
Barbara Feller-Roth

**Proofreader**
Deborah Bruner

**Indexer**
Trisha Feuerstein

**Editorial Assistant**
John Parr

**Composition by**
Laurie A. Steele

**Layout by**
Cindy Putnam

**Production by**
Studio 165

**Separations by**
Color Tech Corporation

**Lithographed in the USA by**
Webcrafters, Inc.

**Consultants**
The author is especially indebted to Michael MacCaskey for his assistance in researching and writing the book.

James R. Baggette, Oregon State University; Albert A. Banadyga, North Carolina State University; Maggie Baylis, San Francisco, Calif.; Russell Beatty, University of California; Louis Berninger, University of Wisconsin; John M. Bridgman, Yorkville, Calif. and Ripton, Vt.; Gerald F. Burke, W. Atlee Burpee Co., Riverside, Calif.; Jack Chandler, St. Helena, Calif.; Andrew A. Duncan, University of Minnesota; Eldridge Freeborn, Atlanta, Ga.; James T. Garett, Mississippi State University; A. E. Griffiths (retired), University of Rhode Island; Anton S. Horn (retired), University of Idaho; Charles A. McClurg, University of Maryland; N. S. Mansour, Oregon State University; John Matthias, San Rafael, Calif.; Fred Peterson, Soil and Plant Laboratory, Santa Clara, Calif.; Victor Pinckney, Jr., Fallbrook, Calif.; Bernard L. Pollack, Rutgers State University; Kenneth Relyea, Farmer Seed & Nursery, Faribault, Minn.; R. R. Rothenberger, University of Missouri; Raymond Sheldrake (retired), Cornell University; W. L. Sims, University of California; Perry M. Smith, Auburn University; William Titus, Nassau County, N.Y.; Doris Tuinstra, Grand Rapids, Mich.; James Waltrip, Gurney Seed & Nursery Co., Yankton, S.Dak.; Frits Went, Desert Research Institute, University of Nevada; Charles Wilson, Harris Seeds, Rochester, N.Y.

**Photographers**
Names of photographers are followed by the page numbers on which their work appears. R=right, C=center, L=left, T=top, B=bottom.

All America Selections: 8, back cover BL
William C. Aplin: 80R, 90L, 126L, 126R, 127R
Liz Ball: 4, 30B, 56, 123L
Josephine Coatsworth: 86R
Thomas E. Eltzroth: 88L, 107L, 111L, 111R, 113R, 120R, 121R, 135L
Derek Fell: 49, 72L, 73R, 75R, 76L, 81R, 82R, 83L, 84R, 87R, 94R, 95L, 97L, 97R, 101L, 101R, 103L, 106R, 113L, 116L, 116R, 120L, 122R, 128, 129R, 130L, 132L
Saxon Holt: 21L, 24, 27, 29B, 46, 47T, 58, 71L
Deborah Jones: Front cover
Michael Landis: 1, 12, 13, 14T, 18, 19, 21R, 31, 32R, 50, 53, 60, 62, 82L, 89R, 98R, 109R, 133R, back cover BR
Robert E. Lyons: 22, 131R
Scott Millard: 32L
Ortho Information Services: 7, 9, 14B, 15, 16, 17, 20L, 20R, 23, 26, 28T, 28B, 29T, 30T, 36, 37, 38, 39T, 39B, 40, 42T, 42B, 47B, 48, 59, 61, 68, 71R, 73L, 74L, 75L, 76R, 77L, 77R, 78R, 79L, 79R, 80L, 81L, 83R, 84L, 85L, 85R, 86L, 87L, 89L, 90R, 91L, 91R, 92L, 92R, 93L, 93R, 94L, 95R, 98L, 99L, 99R, 100L, 100R, 103R, 104R, 105L, 106L, 108L, 109L, 110L, 112L, 112R, 114L, 117R, 118L, 118R, 119R, 122L, 124R, 125L, 125R, 127L, 131L, 132R, 133L, back cover TL, back cover TR
J. Parker: 88R, 129L
Pam Peirce: 70L, 74R, 102L, 105R, 114R, 115L, 115R, 117L, 121L, 124L
PHOTO/NATS: Ivan Masser, 6; Julie O'Neill, 10; David M. Stone, 65, 72R; Marilyn Woods, 78L; Virginia Twinam-Smith, 96L; John A. Lynch, 96R; Ann Reilly, 102R, 108R, 110R, 130R, 134R, 136L, 136R; Gay Bumgarner, 135R
VALAN PHOTOS: V. Wilkinson, 57, 134L; J. R. Page, 70R; Phil Norton, 104L; Val Whelan, 107R; Wouterloot-Gregoire, 119L, 123R


**Front Cover**
Vegetables you raise yourself are always fresher than those you buy, and your selection is as wide as the world's seed catalogs.

**Title Page**
The gardener's reward: This midsummer vegetable and herb harvest shows the skill of the gardener.

**Back Cover**
**Top left:** Beans can be trained to a wide variety of supports.

**Top right:** This simple lean-to protects pepper transplants from cool spring weather.

**Bottom left:** Sugar snap peas have become a garden favorite.

**Bottom right:** Even potatoes can be grown in containers. This nursery can yields a bumper crop.

# All About Vegetables

**WELCOME TO THE GARDEN!**

A general overview of gardening is presented, including some tips to help avoid problems and information on garden planning.

*5*

**THE FUNDAMENTALS OF GROWING VEGETABLES**

The basics of growing vegetables are explored here: how to water, prepare the soil, and protect plants from insects and the elements.

*11*

**PLANNING THE VEGETABLE GARDEN**

The most important step in gardening is deciding what to plant and when to plant it. This chapter shows you how to coordinate your garden with your family's needs and likes.

*41*

**PLANTING AND HARVESTING**

Here is the information you need to get your plants off to a good start, including planting charts, tips on starting seeds indoors, and advice on protecting transplants.

*51*

**THE VEGETABLES**

This chapter gives details about all the common vegetables and many uncommon ones, including how to cook them. Information on gardening methods, planting, avoiding problems, and the best varieties to grow is also provided.

*69*

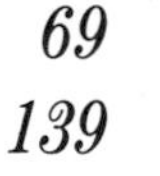

**Seed Sources** *139*

# Welcome to the Garden!

*Vegetables really grow themselves. All the gardener does is help them out a little so they can be their best. Here are a few conditions vegetables prefer and some ways that you can help them.*

There are many answers to the question about why to grow vegetables, maybe as many answers as there are vegetable gardens. No single answer is the right one for everybody. Some gardeners grow vegetables to avoid paying high store prices, or to get better vegetables than the same amount of money will buy at the market—fresher, tastier, picked at the peak of perfection, and served sometimes within minutes.

By growing your own vegetables, you may be able to have a greater selection than is available at local stores, since you can grow types that don't ship well or that may not be in wide enough demand to be readily available at the store. Many home gardeners are freezing, canning, drying, or storing their crops, making vegetables from the garden a joy long after the garden is put to bed for the winter.

Many people who garden feel that getting back to nature helps them to relieve stress. In the garden they can work off their frustrations, have quiet times to think, and make peace with themselves. Growing vegetables may bring a sense of pride, that good feeling when you can pick the first or the largest tomato in the neighborhood, or show off your harvest at the county fair or garden club show.

*A vegetable garden needn't be an unsightly jumble relegated to the backyard. This plot, in a prominent location in the landscape, is as attractive as it is productive.*

*A well-planned vegetable garden will supply the entire family—and perhaps a neighbor or two—with fresh salad fixings throughout the growing season.*

A vegetable garden can bring the family and the community closer together while they work, and later while they eat.

For most gardeners the reason for growing vegetables is the sheer joy and pleasure gained from it. Experimenting with new vegetables and new varieties is as exciting to some as a Sunday afternoon softball game. It is a joy to harvest and think "I did it myself."

Growing vegetables doesn't have to be complicated. Without ever reading a book (even this one), you could easily go to the local garden center and buy some seeds and fertilizer. Then, choosing the sunniest spot in your garden, you could spread some fertilizer on the soil and turn it over with a spade. Then you could plant, following the directions on the back of the seed packets. With just a little bit of luck, you could have a vigorous vegetable garden.

Billions of pounds of beautiful vegetables have been grown by millions of gardeners in just this way, using a marriage of trial and error and common sense. These two factors can go a long way, and this book has no intention of eradicating or complicating them. Instead, the ideas and suggestions offered here should complement your basic horse sense by helping you avoid some pitfalls, obtain good, practical knowledge, and enliven your gardening experience.

If you want to skip the background and get right down to the basic how-to-plant information, turn to the planting chart on pages 63 to 65. This chart will not only help you plant your garden but also help you plan it.

Use the chart for information on how to plant as well as when to plant (at least it will keep you from planting peas and beans on the same day; check the columns, Needs Cool Soil, Tolerates Cool Soil, and Needs Warm Soil). For more specific advice on planting dates and special treatments, check the individual vegetable listings starting on page 69.

## AVOIDING DISAPPOINTMENTS

Most gardeners have succeeded more often than they have failed. When vegetables are grown under the right conditions, the harvest is more likely than not to be bountiful. This is the key to gardening success. To share the successes and avoid the disappointments that other gardeners have experienced, read about the growing conditions that vegetables need, and how to control the factors that can limit success, such as poor soil, improper watering, and insects and diseases. Also note the heading Beginners' Mistakes as you read about the individual vegetables.

## PLANNING THE GARDEN

There's no single right way to grow a vegetable garden. The choice of what and where to plant is a highly personal one, reflecting the interests, knowledge, and imagination of the gardener. You may want your garden to be purely practical, or beautiful, or a mixture of both. Some gardeners plant only the most reliable, success-guaranteed performers, carefully laying out their plots to maximize production. Others remember with fascination the unstructured beauty of grandma's garden, and base their plans on that recollection. And some gardeners are full of surprises, always changing their gardens as their own whims change.

Vegetable gardening is not quite the same as growing vegetables. Vegetable gardening implies straight rows and an orderly sequence of operations: planning for space, choosing varieties, figuring planting dates, and anticipating harvests, among others. It can be a challenging, even awe-inspiring, exercise, especially for the first-timer.

Growing vegetables, on the other hand, means that you can squeeze your plants into any free space, set them among flowers and ornamentals, or just plant them for their own special kind of beauty. Imagine a staggered border of blue lobelia and 'Tiny Tim' alyssum; behind that, a row of cabbage with blocks of 'Thumbelina' zinnias between the heads, or a row of marigolds between rows of potatoes for a good contrast near harvest time. 'Salad Bowl' and 'Ruby' lettuce are beautiful with Iceland

*Vegetable gardening leaves plenty of room for imagination and experimentation. Here, cabbage and stock flourish together in the same planting bed.*

poppies. Or you might consider red chard with alyssum or vinca.

Whether you're interested in vegetable gardening or growing vegetables, and whether your garden site is as small as a bread box or as big as a house, pages 41 to 49 will help you plan your garden.

## INTERPRETING ADVICE

Pages 69 to 138 tell you how to grow and care for each vegetable. You'll learn its use in gardening, its particular requirements, some of its possible uses after harvest, and special tips for growing it to perfection.

You don't have to follow these instructions to the letter; there are no perfect formulas for growing vegetables. However, this book does provide guidelines by presenting throughout the text and in the charts precise temperatures, and measurements for planting, applying fertilizers, adding soil amendments, and other operations.

Don't rely totally on formulas; the garden has too many variables and unknown factors. Many gardeners who deliberately or accidentally deviate from the formulas still grow perfect vegetables—proof that all measurements are not so critical, after all.

However, this doesn't mean that you should discard all gardening advice and measure by instinct. Although it's true that many gardeners let the plants' color, leaf size, or rate of

*'Sugar Snap', the granddaddy of snap peas, is an all-time All-America Selection (AAS) vegetable winner. Many gardeners consider a vegetable patch to be incomplete without this plant, which bears a heavy crop of edible pods on 6-foot vines.*

growth tell them when to fertilize, then measure the fertilizer by the handful or according to how it looks scattered on the soil, these methods are risky. If you wait for the plant to tell you when to fertilize, by then it may be too late to do any good. The biggest—and most frustrating—problem with measuring by instinct is that, should you get exceptional results, you have no sure way of duplicating them in following years.

How important, then, are gardening directions? And how important is instinct? The answer is that both are of value but mostly when combined with another crucial factor: experience. Use the instructions and the measurements (dates, pounds, and inches) as reference points—places to begin and revisit as needed—but make all the adjustments necessary for your own climate and soil. Experience is what will help you make those adjustments; and you get experience only by actually growing the plants yourself, in your own garden with its own conditions. Once you have gotten your brain working, your hands dirty, and your enthusiasm sparked, you can begin to consider yourself a good gardener.

## BETTER THAN THE MARKET

It's well known that homegrown vegetables are superior to store-bought ones, especially corn, snap beans, peas, and all those that lose their fresh quality soon after picking. This is because vegetables that must be picked before they are fully ripe to allow for shipping time lack the taste and quality of ripe, freshly picked produce.

But homegrown vegetables are not necessarily superior. All factors must come together in the right way; and only when the best varieties are planted, given the correct amount of water and fertilizers, kept free of insects, diseases, and weeds, and harvested at the optimum time will the taste be something you will want to brag about: "So much better than that store-bought stuff."

## THE EXTENDED GARDEN

Gardening books often talk about plants that supposedly can't be grown in your climate. For the most part this advice is accurate, but there are many determined gardeners who take great pleasure in growing what supposedly can't be grown. If your interest in the plant is strong enough, you can grow it—even if you have to create a special climate around it.

By using some tricks for creating special climates, you can extend the growing season for weeks in each direction, getting ripe vegetables much earlier and much later in the year than you otherwise would.

Starting on page 24, you'll find information on how to alter unfavorable environments, how to work with nature's whims, and how to swing with nature's rhythms.

You can extend your gardening in another way, too. Each year, more unusual vegetables are made available by seed companies. Try a few unusual vegetables and new varieties of familiar vegetables. You'll find instructions for growing many unfamiliar vegetables beginning on page 69. If you don't know what to do with a vegetable after it's harvested, we give you a few tips for preparation and cooking, too. Friends will ask "what's this?" and you'll be proud to tell them not only what it is but that you grew it yourself!

*The home gardener has the opportunity to raise unusual vegetables such as kohlrabi, a member of the cabbage family that looks like an aerial turnip.*

# The Fundamentals of Growing Vegetables

*No matter what you're growing, certain fundamental principles will ensure gardening success: the right location, good soil, a continual and uniform supply of water and nutrients, matching the crop to the climate, and some protection against pests and diseases. This chapter gives you all the basics you need.*

Plants require certain environmental conditions in order to thrive. Their leaves need lots of light, the right range of temperatures, and some protection from wind, insects, and disease spores. The roots need soil that will hold ample water and ample air and assure a continuous supply of plant nutrients. In large measure, being an expert gardener is a matter of knowing how to provide the environment that allows plants to achieve their fullest potential.

Almost without exception vegetables are sun-loving plants. Some root crops and leafy vegetables tolerate a little shade, but they all grow best where they get full sun. Pick a spot where there are at least six hours of sunshine a day, preferably more. Morning sun and afternoon shade is better than morning shade and afternoon sun, especially in hot climates; because the garden will be cooler, there will be less chance of disease.

*A vegetable garden needs regular tending if it is to produce bountiful harvests of appealing, tasty vegetables. This presents no obstacle for most gardeners, who seem to enjoy the day-to-day tasks of nursing along their vegetable patch.*

*One of the most important factors contributing to a productive vegetable garden is good, workable soil. Most vegetables grow best in a soil that is deep, loose, well drained, and fertile.*

## Tools of the Trade

Any gardener who does a lot of gardening will most likely need few if any additional tools to tend the vegetable patch. A few basics are needed.

- ☐ A square-ended spade and a round-ended shovel for digging large holes and double-digging.
- ☐ A flat rake for breaking up large clumps of soil and raking the soil surface level.
- ☐ A hoe for weeding, cultivating, and opening seed furrows.
- ☐ A trowel (or a large kitchen spoon) for transplanting and applying dry fertilizers.
- ☐ A garden hose, watering can, and perhaps a sprinkler system.
- ☐ A sprayer to apply fungicides and insecticides.
- ☐ Scissors or pruning shears for harvesting. A spading fork can be useful in harvesting root crops.
- ☐ A wheelbarrow or a child's small wagon to carry tools, fertilizers, plants, and other items about the garden.
- ☐ A file, sharpening stone, or diamond file to keep tools sharp.

When buying tools, it's best to select well-made ones that will last the life of the garden. Handle them before buying to make sure they feel good in your hand; with shovels and spades, check the length of the handle for comfort. Always keep tools clean to prevent rust, especially when storing them for the winter, and sharpen them as needed to make your work easier.

Light conditions being equal, select a spot where the soil is good, drains well, and does not have competitive tree roots growing in it. Choose areas where there will not be shade from buildings, fences, and walls, at least in the morning. Make sure that water is available close by. The closer to the kitchen the garden is located, the more convenient it will be.

If you can, choose the south slope of a gentle hill; it will warm up faster in the spring and stay warmer on cold nights as the cold air drains into the valleys. The south side of a wall also warms up more quickly in the spring. However, plants that don't tolerate heat, such as lettuce, won't like that location in the summer. Find a cooler spot for them.

## SOIL

If you are one of those lucky gardeners whose soil is deep, fertile, easy to work, and easy to manage, you will have little to do to improve it. However, most home gardeners are not this fortunate; many have problem soils.

Plant roots need air, water, and fertilizer available to them all the time. The soil makes up the underground environment of the plant and supplies the roots with the elements they need. Problem soils make harsh environments for plants, depriving them of either water or air. However, almost any problem soil can be improved and most can be made into excellent garden soils.

### Air and Water in the Soil

One of the most important factors determining plant growth is how much air is in the soil. Roots breathe, just as people do, taking in oxygen and giving off carbon dioxide. Root tips are constantly growing and need a continuous supply of oxygen for this growth. If they are deprived of oxygen for even a few hours, they begin to suffer and can even die in a day or so.

Two factors—drainage and aeration—depend on the spaces between the soil particles. If the spaces are ideal, they hold ample water after gravity has pulled all the water it can through the soil, and also hold lots of air. If the spaces are too large, the soil drains quickly but doesn't hold much water. Such a soil is said to be droughty, and needs frequent watering. Very sandy soils are often droughty. If the spaces are too small, they remain full of water for a long time after a rainfall and there is no room for air. This type of soil has poor drainage. Some clay soils are in this category. The ideal soil is 50 percent solid material, 25 percent air spaces, and 25 percent water.

For the best plant growth, the soil should remain open and porous from the surface all through the rooting depth. When a soil crusts over after a hard rain, it partially cuts off the air supply to the roots. When the top layer of soil is compacted, the soil underneath suffers for lack of its full requirement of air. Thus, a well-worn path across a lawn will actually kill the grass. Compacted soil not only reduces the supply of air to the roots, it also can limit the amount of moisture available to the plant, since water runs off the surface soil instead of soaking in.

## Improving the Soil

Almost any soil benefits from the addition of organic matter. If you routinely add some sort of organic amendments every time you turn over the soil or plant something new, the soil will improve each year.

Whenever you mix organic matter with fine-textured soil, the soil becomes more mellow and easier to work. Organic matter—for example, compost, peat moss, manure, sawdust, or ground bark—opens up fine-textured soils by separating the soil particles, allowing water to drain from the soil and air to move more readily into it. In sandy soils organic matter acts like a sponge, holding moisture and nutrients in the root zone.

Add enough organic matter to change the physical structure of the soil. This means that at least a quarter to a third of the final mix should be organic matter. If you spread the amendment between 2 and 4 inches deep and work it in to a depth of 6 or 8 inches, you will have the desired result. There is no point in adding just a little dab of anything—minuscule amounts won't change the soil structure. A little peat moss or straw mixed with compressed clay soil will only make an adobe brick.

If you add an amendment that is deficient in nitrogen, such as sawdust, ground bark, or straw, add a generous amount of high-nitrogen fertilizer along with it to help it decompose. Otherwise, the microorganisms that digest the organic amendment will tie up all the nitrogen in the soil until the amendment is largely decomposed.

If you have clay soil and you live in an arid region or an area with salty irrigation water, your soil may contain too much sodium. If so, the soil can be loosened by incorporating some gypsum. Test whether gypsum will help by digging a shallow hole, filling it with water, and timing how long it takes to drain. Then incorporate a cup of gypsum in the soil in the bottom of the hole and fill the hole with water again. The gypsum acts very rapidly; if it has helped your soil, the water will drain much more quickly this time.

## Working the Soil

Turn the soil over with a spade or fork to a depth of 6 to 8 inches, breaking up large clumps and removing stones and debris. A rotary tiller will make the job easier. If the soil is hard and being tilled for the first time, rent the largest rotary tiller you can fit in the

*The best way to improve soil is to dig in lots of organic matter, such as compost or finely ground bark, every year. When using manure or other amendments with a high salt content, add smaller amounts and leach the soil before planting.*

*Top: If the planting area is large, you may want to use a rotary tiller to incorporate organic matter into the soil. Select heavier tillers for harder soil. Bottom: A homemade implement fashioned from scrap lumber does a good job of making the soil level before planting.*

garden; the weight of the bigger machine makes the work easier. If the soil is light or has been turned over recently, it is easy to till with a fork or light tiller.

Before working the soil in the spring, be sure that it has dried out enough. Soil is too wet to work if clods don't break apart easily when hit with the back of a shovel. Another test for moisture content is to take a handful of soil and squeeze it into a ball. If the ball breaks apart when you prod it with a finger, the soil is dry enough to work. If the ball just dents, the soil is still too wet to work; wait a few days and try again. If you work soil too soon, it will form clods that will be difficult to break up.

One trick to working clay soil is to dig when the soil has just the right moisture content. If the clay is still plastic, as described above, it's too wet. Wait another couple of days. If it is hard to penetrate with a shovel and clods are as hard as rocks, it's too dry. Water it deeply and let it drain and dry for about three days, then try again.

One method used to improve soil is called double-digging, which means digging the soil two spade depths deep. To do this, remove the soil one spade depth along an entire row (called a spit). Save the soil in a wheelbarrow or stack it at the other end of the garden for now. Loosen or turn over the soil in the bottom of the spit. You can incorporate organic matter if you wish. Then dig a second spit next to the first, filling the first spit with the soil you remove, again incorporating organic matter as you work. Turn over the soil in the bottom of the second spit and continue this way to the far end of the garden. Use the soil from the first spit to fill the last.

Double-digging is a great deal of work. Try it in one bed or section of the garden and see if your effort is rewarded with an improvement in plant vigor before you tackle the entire garden. It isn't necessary to double-dig every year. The effect of one double-digging will last for several years.

After the soil has been turned over, rake it level and remove surface clods, but don't rake it enough to turn it into dust. If you do, it will hold too much water and develop a crust after the first rain or watering. The surface texture should be finely granular; it should be like sugar instead of flour.

*This set of three compost bins has removable front boards to make turning easy.*

### Compost

The black, fragrant, crumbly, partially digested organic residue called compost comes from garden waste material. Compost is an excellent source of organic material for improving the soil. Whatever method of composting you use, the main objective is to arrange the waste material in such a way that the soil organisms that break down the waste can thrive and multiply. These organisms need moisture, air circulation, and food.

To build a compost pile, use a mixture of green and dry materials. Green material decomposes rapidly; grass clippings, lettuce leaves, pea vines, and other succulent materials contain sugar and proteins that provide excellent nutrients for the soil organisms. On the other hand, dry material—sawdust, dry leaves, small twigs, and prunings—contains very little nitrogen and decomposes very slowly when composted alone. By mixing green and dry materials, your pile will compost at just the right rate.

The size of the composting particles will affect the rate of decomposition. If dry leaves and other dry materials are put through a shredder rather than added to the pile as is, the small pieces will decompose faster (more surfaces are exposed to the decay organisms). Shredding also creates a fluffier mixture, making air and water penetration more efficient. You can buy or rent a shredder or you can simply use a rotary mower to shred leaves. If you produce grass clippings in large quantities, mix them thoroughly into the composting material; otherwise, they will form a soggy mass, putrefy, produce unpleasant odors, and attract flies. After you've mixed the grass into the compost, spread a layer of soil or old compost over the top as an insulating layer.

For the sake of convenience, divide the compost area into space for three piles. The first is for the daily collection of waste products—refuse from the vegetable harvest, fruit and vegetable wastes from the kitchen, coffee grounds, eggshells, small prunings, and so on. The second pile is for the fast-working compost; add nothing to this pile. The third pile is to store the finished compost.

When enough material has accumulated in the first pile, build the second pile in layers, adding some manure, soil, or old compost to each layer to provide decomposition bacteria. Wet down dry materials as you add them. In a few days the center of the compost pile will heat up to as much as 160° F. Take its temperature by sticking a pipe or crowbar into its heart for a few minutes, then removing it to see how hot it is.

After a couple of weeks, the temperature will begin dropping. When the pile is no longer hot but just warm, turn the pile by shoveling it into a new pile or an empty bin. As you shovel it, put the outside of the old pile in the center of the new pile. If the compost is dry or only slightly damp, water it as you turn it. Turning mixes and aerates the parts thoroughly. The new pile will again heat up but not as much as before. When it cools again after a few weeks, it is ready to use.

*A favorite mulch in the vegetable garden, straw is usually tilled in after harvest to improve the soil.*

### Soil Acidity and Alkalinity

Acidity and alkalinity are measured in a unit called pH; these values range from 0 to 14, with 7 representing neutrality. Numbers less than 7 indicate increasing acidity, and numbers greater than 7 indicate increasing alkalinity.

For optimum development most vegetables require a pH level of 6.0 to 7.5. (Potatoes are one exception—they prefer a more acidic soil.) If soil is too acid, raise the pH with lime; if the soil is too alkaline, lower the pH by adding some soil sulfur.

If lawns in your area receive yearly liming treatments, you can be sure that your own soil is also naturally lime deficient. This is true for gardeners living in Mississippi and eastward in the South, or in Ohio and eastward in the North. In western areas where annual rainfall is low, alkaline soils tend to be the general rule.

There are several types of lime available to raise pH. Hydrated lime, which is quick acting, should be applied several weeks prior to planting and watered in well to avoid any likelihood of burning. Crushed limestone is much slower acting and longer lasting in its effect. It requires a heavier application but can be used with less chance of burning. Dolomitic limestone is particularly good, because it contains the trace element magnesium.

### Soil Tests

Soils may be ideal for one kind of plant but less than ideal for another. A soil in which rhododendrons and azaleas will thrive will not support tomatoes, for example. To grow the best vegetables you can, it's worthwhile having your soil tested.

If there is any question about the composition of the soil in the garden, it is wise to have it tested and not guess how much the pH needs correcting or how much fertilizer needs to be added. Many states offer free or low-cost soil tests. Take advantage of this good investment and call the local county extension agent for details. Soil test kits that measure pH can be purchased at a local garden center. There are also private laboratories that can run full soil analyses, which include, in addition to the pH, amounts of organic matter and the essential elements for nutrition that are present.

### Mulches

Adding mulches offers a variety of benefits: Mulches protect roots in the top inches of soil from high temperatures; they conserve water by reducing evaporation from the soil; and they prevent erosion and reduce soil compaction caused by foot traffic or by water from heavy rains or sprinklers. They reduce the number of weed seeds that germinate, and a summer organic mulch, tilled back into the soil in the fall after harvest (or the next spring before planting), will improve the soil.

On the negative side, mulches can also harbor insect pests. Most of the insects that thrive under a mulch are beneficial or neutral, but some can cause plant problems.

Organic mulches and soil amendments may be made of the same materials; how they differ has to do with when and where the application is made. A mulch covers the surface of the soil; it is not worked in. Applied in early spring it will keep the ground cool; applied in late spring, after the soil has warmed up, it will keep the soil warm.

Mulches may be made of organic material (such as leaves, straw, manure, sawdust, ground bark, compost, and the like) or manufactured materials (such as black or clear polyethylene film, burlap, aluminum foil, or paper).

Peat moss does not make a satisfactory mulch. It blows when dry and is difficult to rewet after it has dried out.

For established plants apply organic mulches 1 to 2 inches thick for fine materials (such as sawdust), and up to 4 inches thick for coarse materials (such as straw).

Film mulches, such as black plastic or treated paper, are excellent. Cover the ground with the film before planting and weight down the edges with soil. Cut slits or *X*'s in the film to plant through.

### Synthetic Soils

If you'll be growing vegetables in containers, your best bet is a quality-controlled, synthetic soil mix, usually referred to as planting mix. Planting mixes can also be used with great success in raised beds.

The organic part of the mix may be peat moss, redwood sawdust, shavings, hardwood bark, fir bark, pine bark, or a combination of any two or more.

The mineral part of the planting mix may be vermiculite, perlite, pumice, coarse sand, or a combination of two or more. The most commonly used minerals are vermiculite, perlite, and coarse sand.

When mined, vermiculite resembles mica. When vermiculite is heat treated, air spaces form to expand its mineral flakes up to twenty times their original thickness.

Perlite is a granitelike volcanic material, which is mined. When it is crushed and heat treated, it pops like popcorn and expands to twenty times its original volume.

The planting mix you buy may be 50 percent peat moss and 50 percent vermiculite, 50 percent wood products and 50 percent coarse sand, or some other combination of organic and mineral components. But although the ingredients may vary, the principle behind all mixes remains the same—soilless "soil" must provide the following.

*Redwood bark forms neat pathways among these narrow raised beds of strawberry plants. The black plastic mulch in the beds hastens fruiting in addition to discouraging weeds.*

*When growing vegetables in containers, ensure good results by using a quality-controlled synthetic soil mix. Here, a bumper crop of red potatoes is harvested from a small plastic trash barrel.*

☐ Fast drainage of water through the soil.
☐ Air in the soil after drainage.
☐ A reservoir of water in the soil after drainage.

The most important element in any mix intended for containers is how much air is left in the soil after drainage. Container soil must drain better than garden soil. In the garden, water moves through a column of soil with continuous capillary action (like a blotter). But in a container, that continuity is broken by the existence of the container bottom, gravel, or whatever impedes the water's downward flow; where the continuity is broken, the water builds up.

Planting mixes circumvent this problem. When the mix is watered, the water drains quickly from the macropores (large pores), thus allowing air to follow the water down to the roots. At the same time, however, some of the water is retained in the micropores (small pores) of the mix.

Soil mixes are ready to use as is. Just moisten the mix, fill the container with it, set in the transplant, water, and watch it grow.

However, feel free to tamper with your store-bought mix, if you choose. You can add sand (10 percent by volume) to weight down the mix, which is especially useful if small containers are to be placed in windy spots, or you can use perlite to lighten a mix that tends to be too heavy.

Keep it simple: Some gardeners like to work out complicated mixes containing five or six ingredients, but this isn't necessary. A simple combination of peat moss and perlite or coarse sand gives good results.

If you're planning to use many containers or fill a few raised beds, it's more practical to make your own mix than to buy a ready-made one. Here is a recipe for 1 cubic yard (27 cubic feet) of such a mix.

9 cubic feet of coarse sand
18 cubic feet of ground bark or nitrogen-stabilized sawdust

or

9 cubic feet of coarse sand
9 cubic feet of peat moss
9 cubic feet of ground bark

To either of the above formulas, add:
5 pounds of 5-10-10 fertilizer
7 pounds of finely ground dolomitic limestone
1 pound of iron sulfate

## WATER

Garden soil is a reservoir for water and plant nutrients. Your job, as caretaker, is to keep the reservoir healthy (as explained in the preceding section) and full of water. There are many ways of recharging the reservoir, rain being the primary method over most of the country. You can also use a watering can, a drip irrigation system, a sprinkler system, a handheld hose, or many other systems.

With most systems the soil reservoir dries out to a certain extent, then is refilled to its capacity when you water. If you water too often, it never gets a chance to dry out and stays too wet. This is called overwatering. (Watering for too long at a time isn't overwatering, just wasting water. The excess drains away.)

## *Critical Period for Watering*

Flowers may fully recover from the retarded growth caused by water stress, but vegetables will not. For a successful harvest, make sure that these vegetables receive ample water during the following critical periods.

| Vegetable | Critical Period |
|---|---|
| Asparagus | Fern |
| Broccoli | Head development |
| Cabbage | Head development |
| Carrot | Root enlargement |
| Cauliflower | Head development |
| Corn | Ear silk and tassel development |
| Cucumber | Flowering and fruit development |
| Eggplant | Flowering and fruit development |
| Lettuce | Head development |
| Lima bean | Pollination and pod development |
| Melon | Flowering and fruit development |
| Onion | Bulb enlargement |

You can tell when to water by feeling the soil a couple of inches below the surface. If it feels cool to the touch but doesn't wet or muddy your finger, it's just the right time to water. If it's dry and dusty, you've waited too long. If it wets your finger, it's too soon to water.

Soils on the clay side have a high water-holding capacity; the air spaces are very small, allowing water to move through them slowly. Such soils need to be watered less frequently than sandy soils, but slowly and for a long time. Sandy soils hold only a little water. They should be watered more frequently but for only a short time.

### Underwatering

If you don't water often enough, the plants will be under water stress—a malady that progresses from slight to severe. A thirsty plant must work harder as the moisture supply in the soil decreases.

Many plants have a remarkable ability to recover from water stress. When impatiens, for example, suffers from lack of moisture, it drops its flowers, droops down, and appears to be dead. However, one lavish application of water should be sufficient to make it straighten up within only a few minutes. Soon afterward, it will begin to develop a new crop of flowers.

Occasionally, gardeners use water stress as a deliberate tactic—for example, when growing herbs for a seed crop. And many flower-producing annuals respond to water stress by flowering even more profusely. To the annual plant, a dry spell is a signal to get busy with the business of flowering and setting seed before it's too late.

However, water stress rarely pays off in the vegetable garden except in unusual cases—for instance, if a tomato plant is producing only vines, it will grow and ripen fruits when put under water stress. But for the most part, putting vegetables under stress will exact a

*Vegetables need a steady supply of water, especially when they are flowering or fruiting. Since hand watering is inexact, checking soil moisture regularly is particularly important for gardeners who irrigate by hand.*

severe penalty. Snap beans will drop their blossoms. Lettuce—whose shallow root system requires a steady supply of moisture—will turn bitter. Cucumbers will stop growing altogether (although they will resume if watered again). Beets will become tough and stringy. Radishes will turn hot. Turnips will develop too strong a flavor. Muskmelons will lose their sweetness.

### Proper Watering

There are no exact formulas for watering, because there can be variations in soil, temperature, and drying breezes. The basic rule of thumb is to water thoroughly, filling the root zone. Then let the soil dry out a bit, and when it needs it, water again.

How can you tell what thorough watering means to your specific garden? Apply what you think is a sufficient amount of water. Then, with a spade or shovel dig down about a foot and take a look. Did the water penetrate this far? If yes, you are watering properly. If no, you need to water longer.

In average soils, between 1 and 2 inches of water per week should be applied. Measure rainfall with a rain gauge and compensate for the difference when watering.

You can develop a feel for how long to wait between waterings by observing the appearance of the soil and using your hands to feel it.

As to when during the day to water, anytime will do. However, the early morning offers two advantages: The plants will lose less water to evaporation, and the plants will stay dry at night and thus be less susceptible to attack by disease-causing organisms.

When watering plants in containers, keep applying the water until it drains out the bottom, which will leach nitrogen and potassium through the soil mix. This may seem like a waste of fertilizer, but it offers a distinct advantage by preventing harmful salts from building up in the mix.

### Furrow Irrigation

Gardeners who have a variety of vegetables crowded into a small area frequently prefer furrow irrigation. Overhead sprinklers cannot be selective enough to avoid those vegetables that do not like wet leaves.

A furrow is a shallow ditch, dug alongside each vegetable row, into which water is applied. Furrow irrigation confines the water to the plant's root zone and also inhibits weed growth in unwatered areas.

When you lay out a furrow irrigation system, keep the furrows as level as possible. Too much slope will cause the water to flow rapidly and some areas may receive too little water. You can, if you have the room, place the furrows to allow dry walkways between paths.

### Drip Watering

The idea of making a little water go a long way is nothing new, especially in areas with a limited supply of water. What is new, however, is

*Left: Furrow irrigation—flooding the ditches between the rows—is a common way of watering row crops.*
*Right: Drip irrigation encourages healthy growth by delivering water slowly and evenly to the root zone.*

some equipment and irrigation techniques. Drip or trickle irrigation systems use less water than other watering techniques, because the water is applied only where it is needed and there is less loss of water from evaporation and wind.

With drip irrigation, the water drops onto the soil surface without disturbing the soil structure, and the water can sweep between soil particles. Once in the soil, the water moves by capillary action to the surrounding areas.

Drip irrigation drops the water onto the ground through emitters, which are the heart of the system and control the rate at which water drips. Water moves only a foot or so through the soil away from each emitter, so an emitter must be placed at the base of each plant or every foot or so along a row. Other types of emitters leak water through minute holes or a porous wall of a hose.

In drip irrigation one tries to replenish the water in the root zone on an almost daily basis. In other words, drip irrigation does not store water for future plant use; rather it constantly replaces water that has already been used. Instead of making wide swings from too wet to too dry between waterings, the amount of water in the soil remains almost constant. Plants generally grow better with a drip system than with other types of watering.

Drip systems have their problems, however. They must be checked frequently to make sure that they are watering properly and aren't plugged up. Unless the irrigation water is very clean, they clog easily.

The kind of drip system that emits water through a porous hose generally is less trouble than the point-emitter type. Because the porous hose system operates on less pressure than most point-emitter systems, they are not compatible.

Drip systems are usually installed with a timer. The battery-operated type is sufficient for most gardens and almost frees you of the job of watering.

If you want to experiment with a drip system, buy a small kit and try it in one part of the garden for a year.

### Sprinklers

There are two kinds of sprinklers—underground sprinkler systems and hose-end sprinklers. The underground systems are good for large gardens. If installed on a timer, they will water with convenience and regularity. For smaller gardens you can use either underground systems or hose-end sprinklers. The latter will cost less, although they offer less convenience. They come in many sizes and shapes. Choose the one best suited to your area, and one that will use water most efficiently.

### Watering by Hand

Watering by hand is very time-consuming and inefficient unless the garden is a small one. Although some gardeners enjoy watering their

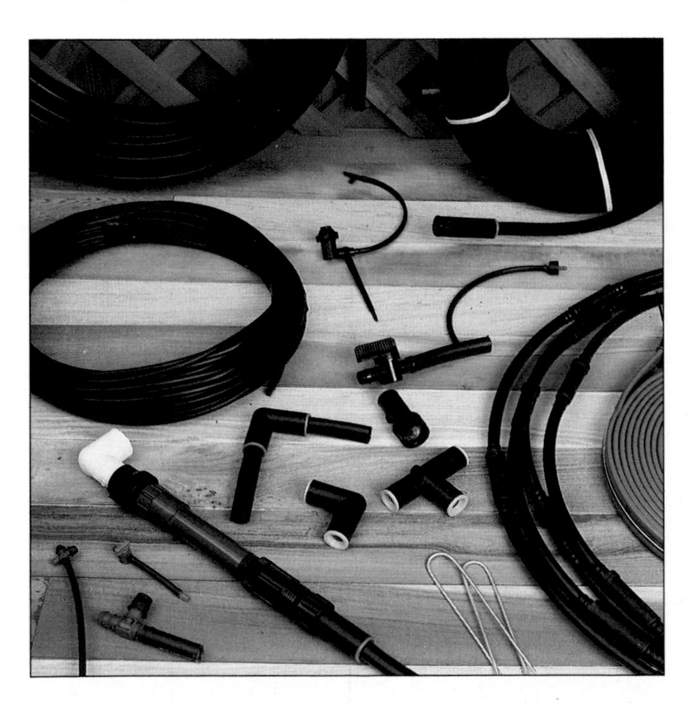

*Left: Drip irrigation is efficient, economical, and easy to install. The components include tubing, connectors, emitters, and valves.*
*Right: This tubing emits a fine spray of water from small holes.*

*Leafy vegetables such as this crop of 'Melody' spinach produce the tastiest results when grown in rich, fertile, moist soil.*

garden and feel that it gives them a chance to look closely at growth and detect any problem signs early, the time spent is considerable and there's the chance of erring on the side of watering too little.

## FERTILIZERS

On a daily basis plants require only a small amount of nutrients, but that amount must be available just when plants need it. A few slow-growing plants allow some leeway—you don't need to fertilize until you see the lower leaves begin to yellow. But vegetables won't let you be this casual. Right away they demand adequate nutrients in the soil to see them through to harvest. An insufficient amount of nutrients will retard growth and this, in turn, will reduce both quality and yield.

### Components of Fertilizer

All commercial fertilizers are labeled with the percentages they contain of nitrogen (N), phosphorus (P), and potassium (K). There are many formulas, but the percentage of nitrogen is always indicated by the first number, phosphorus by the second, and potassium by the third. Nitrogen is the most important element; the percentage of it in the formula dictates how much fertilizer to apply.

Nitrogen is necessary for stem and leaf growth and dark green foliage. Phosphorus is needed for root growth, flower production, and fruit development. Potassium is critical for plant metabolism and food manufacturing. Many fertilizers also contain what are called minor, or trace, elements that are needed for plant health and growth.

For the best production, the ratio of nitrogen, phosphorus, and potassium should be approximately 1-1-1, 1-2-1, or 1-2-2. Commonly available commercial fertilizers with this ratio are 10-10-10, 5-10-5, 6-12-6, and 5-10-10. However, the amounts you apply depend on minerals naturally present in the soil and in the soil amendments you use.

### Organic and Inorganic Fertilizers

Organic fertilizers are those made of animal or vegetable by-products and include fish emulsion, blood meal, bonemeal, alfalfa meal, cottonseed meal, sewage sludge, and animal manures. Some gardeners prefer to use only organic fertilizers because they are slow acting and don't burn, but they are bulky, sometimes unpredictable in their nutrient release, and usually more expensive. Some organic amendments, such as compost and manure, contain enough nutrients to act as a fertilizer when applied as an amendment.

Inorganic fertilizers are mineral compounds, such as potassium nitrate, ammonium nitrate, and ammonium phosphate. They are much faster acting than organic fertilizers and can burn plants if not diluted with a thorough watering after applying. They also leach more quickly from the soil than organic fertilizers.

In general, a combination of organic amendments mixed into the soil before you plant and later side-dressings of inorganic fertilizers give the best results.

### Dry and Liquid Fertilizer

Many fertilizers are soluble in water. Some are granular and are meant to be applied to the soil and dissolved with a thorough watering. Others are very soluble crystals that are dissolved in water and then applied through a hose. Still others are sold as liquids to be applied diluted with water. In general, all do a good job of feeding if you apply the right amount. Choose one that you find most convenient or pleasant to work with.

### Slow-Release Fertilizer

Soluble fertilizers make nutrients immediately available to the plants but leach out of the soil in a few days or weeks. Slow-release fertilizers relinquish nutrients over a period of months. Most organic fertilizers are slow release; as microorganisms decompose the fertilizer, the nutrients are discharged into the soil. Some synthetic fertilizers also release nutrients slowly. Some are only slightly soluble in water; others are coated with a polymer that slows their release. Many commercially prepared fertilizers contain a mix of immediately available and slow-release forms.

## *Some Fertilizer Tips*

Here are a few tips to help you avoid fertilizer problems and get the most value for your effort.

- ☐ Be sparing with all types of fertilizers. Too much manure can cause as much trouble as too much of any other kind of fertilizer. When using manure or other organic sources of nitrogen, cut down on the rate of fertilizer application.
- ☐ Follow label directions.
- ☐ Follow up dry fertilizer applications with a thorough watering to dissolve the fertilizer and carry it into the root zone.
- ☐ When using large amounts of fertilizer for such heavy feeders as cabbage and onions, apply half the amount before planting, then side-dress with the remainder once growth is under way.
- ☐ Check the timing for applying fertilizer for each vegetable. The first application is crucial with many crops, since fertilizer is beneficial to early vigorous leaf growth.
- ☐ Give less nitrogen to plants grown in partial shade than to the same plants grown in full sun.
- ☐ Increase the amount of fertilizer when plants are crowded into narrowly spaced rows and when plants are grown in a random pattern in a small plot.

### How Much Fertilizer?

When you apply fertilizers, whether fish emulsion, blood meal, commercial liquid, or commercial dry fertilizers, be sure to follow the directions on the package. Don't try to outguess the manufacturer. If you must err, err on the side of too little. Too much of any fertilizer, even manure, is dangerous, since it will burn the roots and can cause the buildup of too much salt in the soil.

*When fed and watered generously, ornamental flowering cabbage forms succulent, tender leaves that can either be cooked or used raw in salads.*

**Applying Dry Fertilizers**

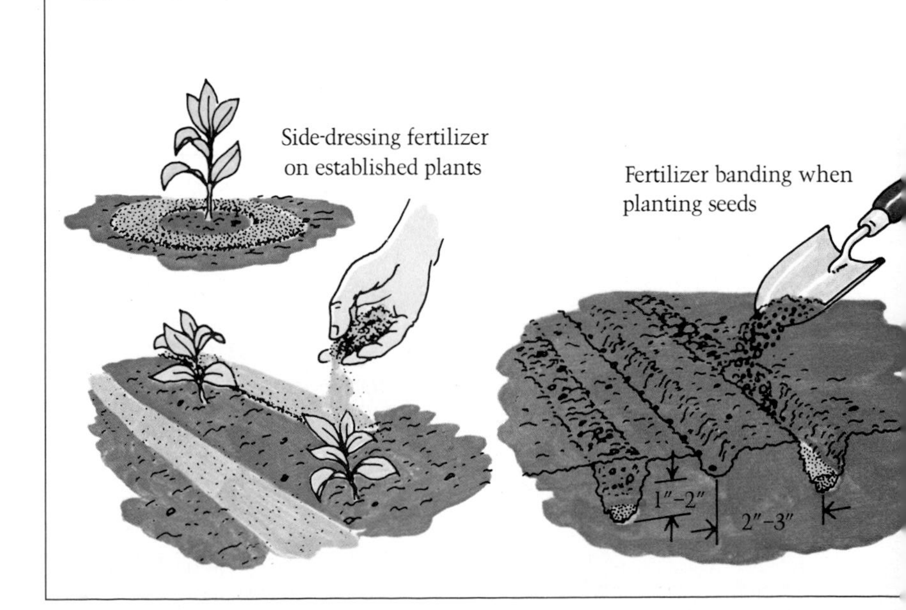

## Fertilizer Amounts

The following chart shows how the amount of fertilizer to be applied decreases as the percentage of nitrogen increases.

Assume that 5-10-10 fertilizer recommendations call for 3 to 4 pounds per 100 square feet.

| Formula | Pounds per 100 Square Feet |
|---|---|
| 5-10-10 | 3.5 |
| 6-20-10 | 2.8 |
| 8-24-8 | 2.0 |
| 10-10-8 | 1.7 |
| 16-16-16 | 1.0 |

If the fertilizer you are using contains one of the higher percentages of nitrogen, use less. Read the directions on the package to find out the nitrogen percentage; otherwise, you may unintentionally double or triple the amount of nitrogen needed.

### Applying Fertilizer

When applying dry fertilizer, it's easy to figure the amount to apply per square foot or length of row if you use liquid measurements. One pint of dry fertilizer weighs about 1 pound, 1 cup weighs ½ pound, and so forth.

Before planting, mix the fertilizer with the soil. Spread it evenly over the soil at the rate called for on the package, then work it in with a spade or tiller.

**Fertilizing container plants** Vegetables need a small but continuous supply of nutrients. When the weather is hot and the garden is watered more frequently, part of the nutrient supply is leached away. A good way to fertilize is with a weak nutrient solution in every watering, taking a cue from your watering schedule rather than from the calendar.

Reapply the fertilizer as a side-dressing after the plants are up and growing. Scatter it on both sides of the row 6 to 8 inches from the plants. Rake it into the soil and water thoroughly. Refer to the descriptions of the individual vegetables you will be growing for fertilizer needs for specific crops.

*This polymer-coated fertilizer releases nutrients slowly to the lettuce.*

## CLIMATE AND MICROCLIMATE

*Climate* is the overall weather pattern in your area: how cold it gets in the winter, when rain falls, how much snow accumulates. *Microclimate* refers to the specific weather conditions where a plant is growing. A spot on the north side of a house that never gets sun has a very different microclimate than one on the south side. You can't do much about the climate in your area, but you can both choose microclimates and change them to suit your plants—sometimes dramatically.

### Climate

For vegetable growing, the most important yardstick is the length of the growing season. Northern and high-altitude gardeners have short growing seasons; only a limited number of vegetables will ripen before a crop is killed by the first fall frost. (On the other hand, since plants are fueled by sunlight, they grow spectacularly fast and large in the long summer days of the northern latitudes of Canada and Alaska.) The length of the growing season in the United States and Canada varies from barely 3 months to as much as 12 months. Contact the local county extension agent to determine the dates of the first fall and last spring frost in your area.

Once you have determined the length of your growing season, refer to the specific vegetables you want to grow in the chapter starting on page 69. The time from planting to harvest is given for each vegetable and, where applicable, for individual varieties.

### Day Length

The length of the day, or more specifically the number of hours of actual sunlight, influences the growth habits of several annual vegetables, particularly onions, spinach, and Chinese cabbage. The lengthening of days beyond the 12-hour day/night cycle of the vernal equinox signals spinach and Chinese cabbage to flower. Temperatures begin to rise, making the environment more favorable for flowering. The term used to describe this is *bolting,* which means that the vegetables will form a flower and set seed before they are ready to be harvested.

It's best to plant when the day length is most favorable for a particular vegetable. Onions are divided into long-day and short-day varieties and will not form bulbs if the variety does not match the day length. When Chinese cabbage is planted during the long days of spring, it rarely grows properly. But when it's planted to mature in the short days of fall, it grows well.

Spinach also requires short days; long days cause bolting, especially if the plants were subjected to cool conditions when small. To avoid bolting, plant in early spring and use a

**Gardening in Tune With Nature**

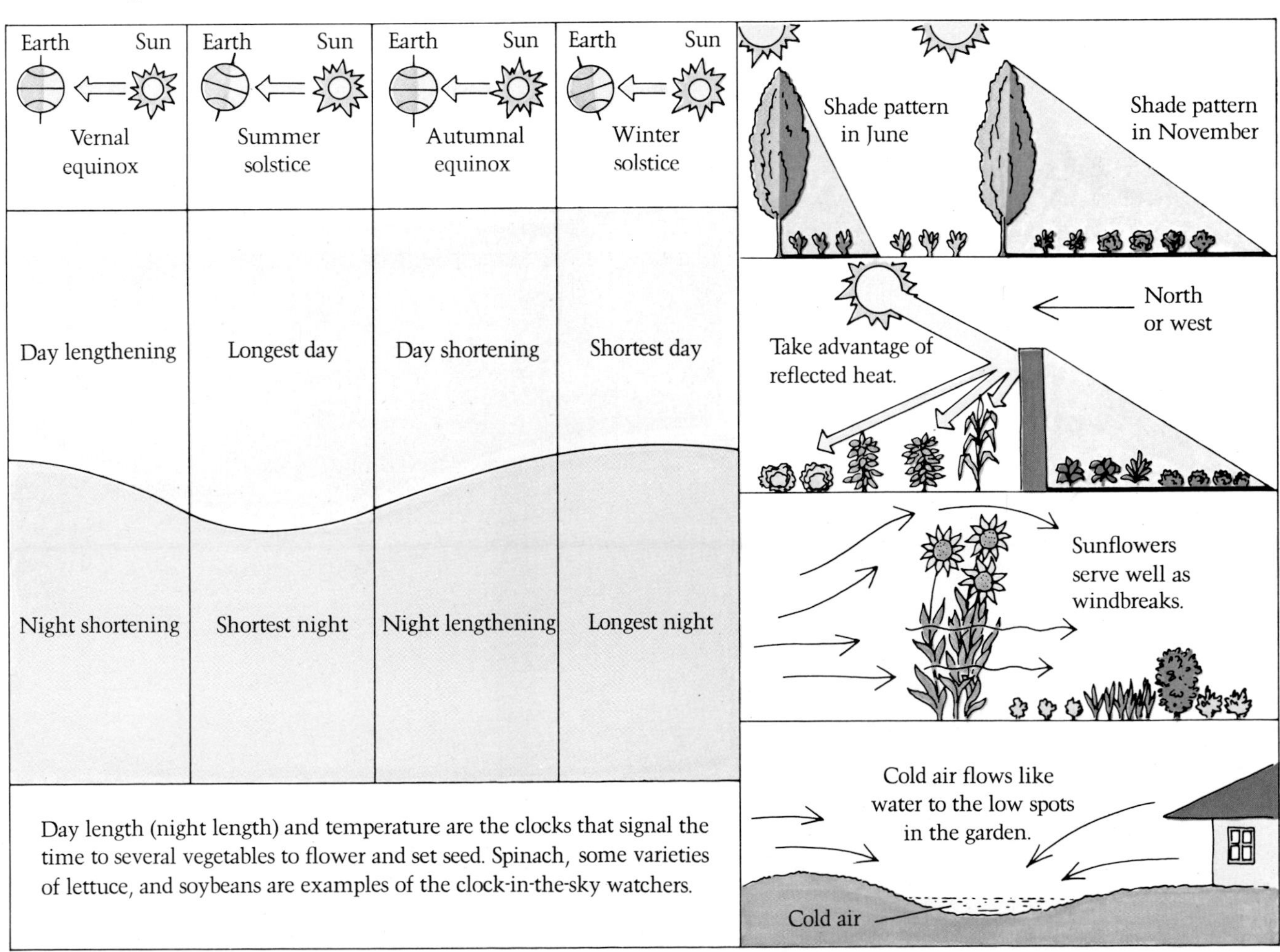


Day length (night length) and temperature are the clocks that signal the time to several vegetables to flower and set seed. Spinach, some varieties of lettuce, and soybeans are examples of the clock-in-the-sky watchers.

*Although many plants are damaged or killed by cold, the flavor of some of the cool-weather vegetables, such as kale, celery, parsnips, and brussels sprouts, is improved by a light frost.*

long-standing variety. Or, in mild-climate areas, plant in the fall.

Breeders of vegetables have applied considerable skill to the problem of bolting. To avoid or reduce the losses caused by bolting, they have created bolt-resistant varieties. Look for them when you are shopping for vegetables that might otherwise tend to bolt.

### Type of Vegetable

Vegetables are divided into cool-season and warm-season classifications. Cool-season vegetables grow best during cool weather; warm-season plants need heat to grow well. Check the planting chart on pages 63 and 64 to find out which vegetables are cool-season and which are warm-season plants.

Cool-season plants should usually be planted before the last frost in the spring; in mild-climate areas they should be planted in the fall to ripen in fall or early spring. If they mature during hot weather, they are often tough and bitter, they do not grow well, and they may bolt, or send up a seed stalk before they reach full size. Bolting destroys their flavor.

Warm-season plants should be planted after the last frost, sometimes after the soil has warmed up, late in the spring. They are killed by frost and need warm nights to mature properly. If the soil temperature is below 50° F,

## *Cool-Season and Warm-Season Vegetables*

Vegetables are commonly divided into cool-season and warm-season plants. This list groups them that way for quick reference. For more specific information about each vegetable, see its entry in the chapter beginning on page 69.

| | |
|---|---|
| Distinctly cool-season crops that prefer 60°–65° F, are intolerant of high summer temperatures above a monthly mean of about 70°–75° F, and tolerate some frost. | Cabbage, kale, broccoli, cauliflower, turnips, rutabagas, kohlrabi, spinach, parsnips, lettuce, and celery |
| Crops adapted to a wide range of temperatures: | |
| Prefer 55°–75° F and tolerate some frost | Onions, beets, garlic, carrots, leeks, shallots, and potatoes |
| Prefer monthly means of 65°–80° F and will not tolerate frost | Muskmelons, cucumbers, squash, pumpkins, beans, tomatoes, peppers, and sweet corn |
| Distinctly warm-weather, long-season crops that prefer a temperature mean of about 70° F and tolerate no cool weather. | Watermelons, sweet potatoes, eggplant, some peppers, and okra |
| Perennial crops | Asparagus, globe artichokes, and rhubarb |

beans will rot and tomatoes and eggplant will sit and sulk. Lima bean seed is likely to rot if the soil is below 62° F; the same goes for okra.

It should be made clear that gardens subjected to cool temperatures all summer can produce marvelous cabbage and lettuce but will not be able to support muskmelons and sweet potatoes. The reverse is true for gardens that receive high temperatures all season.

### Microclimate

You can control your garden's microclimates by understanding a bit about how gardens gain and lose heat.

Heat comes from the sun, which warms the soil. The soil is a reservoir for heat, as it is for water and plant nutrients. At night the ground radiates the heat it absorbed during the day; thus the soil cools. The absorption qualities of various materials can be used to increase heat storage or decrease it, as desired. For instance, the orchardist floods the ground to reduce frost potential; the water holds heat, which slows outgoing radiation.

Mulch is the gardener's favorite insulation material. Loose mulches, such as grass clippings, leaves, or straw, reduce the radiation that reaches the soil and keeps it cooler. For this reason, it's better not to mulch the garden in the spring until the soil has warmed to the temperature you want. A summer mulch keeps the soil cool.

Solid walls and fences around a garden can trap radiation, making the garden warmer than the surrounding areas. Rocky areas, pavement, and masonry can also make hot pockets, catching the heat of the sun and radiating it onto cold areas nearby. Allowing melon vines to trail across an asphalt driveway will keep the fruit warmer at night and make it sweeter.

Color also has an effect on radiation: A light-colored wall will bounce more light to nearby plants than will a dark-colored wall. Using a light-colored wall or constructing a panel painted white or covered with aluminum foil may enable a gardener to produce a crop that might otherwise not have been possible, or to increase its yield.

### Slope of the Land

As the air next to the ground cools at night, it becomes heavier. Like water, it flows slowly downhill into washes, valleys, and basins, filling depressions and low spots as it progresses.

A vegetable garden located at the bottom of a slope will be cooler than one located along or on top of the slope. Slopes can be miniature banana belts or potential frost zones, depending on their direction of pitch. If the slope has a southern exposure, it will be warmer and drier

*Although small, this hot, south-facing plot provides ideal conditions for growing corn and zucchini.*

*Top: When attached to cord strung between posts, clear plastic film becomes an efficient windbreak.*
*Bottom: Portable wood and wire frames protect plants from birds, rabbits, squirrels, and other animal pests. Covered with clear plastic film, the frames serve as miniature greenhouses.*

than one with a northern exposure, affecting the crops that can be grown.

### Wind

The cooling and drying effects of wind on a garden causes water to transpire from leaves quickly and evaporate from the ground more rapidly than in calm weather. Although a gentle breeze may cool a hot garden, the need for watering will be increased. A strong wind can have a detrimental effect on a garden, and some protection must be afforded. If you live in a windy area, erect a lath or fabric wind fence on the upwind side of the garden. Barriers of plants or structures, properly located, will also temper prevailing winds.

## EXTENDING THE SEASON

A determined gardener can always cheat the season by manipulating the microclimate, which is especially helpful for specific crops. They might benefit from a warmer or cooler season, or more or less moisture, or less sun, or less wind. Whatever the need, there are ways to make plants more comfortable.

You can warm the soil earlier in the spring by covering it with clear plastic, which collects heat rapidly. After the soil is warm, remove the plastic (it makes the soil much too hot for plants during warm weather), plant the seeds or transplants, and cover them with a row cover.

### Row Covers

A small greenhouse can be erected over each row of vegetables by using a fabric row cover. Some row covers, such as those made of polyethylene film, need to be supported by hoops of wire or some other material, or by a wooden frame. Others are called floating row covers because they are so lightweight that they can lie on the plants without harming them. Look for a row cover that is ventilated so you don't need to open and close it as the weather changes.

Weight down the edges of the row cover with soil or boards to keep it from blowing. It's difficult to weed under a row cover, so before you put it in place, lay down a mulch to stop weed growth. Black plastic is effective. Some row cover materials allow water to pass through; others don't. If you have a row cover that doesn't permit the passage of water, install a drip irrigation system under it.

Row covers allow you to plant two or more weeks earlier than normal in the spring, and also allow plants to grow later in the fall. They have the added benefit of protecting vegetables from insects and diseases. Plants generally grow more vigorously under row covers during cool weather and are less damaged by insects, wind, and sun. If several gallon bottles filled with water are placed inside the row cover, they will absorb heat during the day and emit it during the night to further protect plants from the cold.

Remove the plastic row cover when the weather gets hot. If you have a wire hoop or other support system, you can replace the plastic with cheesecloth or screening to keep insects away from the plants. If you are growing a cool-season crop, such as lettuce, during the summer, drape a couple layers of cheesecloth or shade cloth on the supports; raise the sides high to allow free air circulation during hot weather.

### Cloches

Individual plants can be protected with a cloche. Developed during the nineteenth century, cloches are bell-shaped jars without bottoms. Today, you can buy paper caps to protect individual plants, or use plastic gallon jugs with the bottoms cut out.

*Top: This portable hinged A-frame covered with clear plastic film warms the soil in spring. The open end permits air circulation.*
*Bottom: Made from lightweight material, a floating row cover rests directly on plants without harming them.*

### Cold Frames and Hotbeds

One of the best ways to extend the growing season is with a cold frame or a hotbed. At the turn of the century, no farm garden was complete without a cold frame—a bottomless, usually glass-covered box that was heated only by the sun. Made of redwood or pressure-treated wood, the box is sunk into the ground, where the soil provides insulation. Its hinged, transparent window can be made of stock cold frame sash (available from greenhouse supply firms), an old window or a glass door, fiberglass, or polyethylene film. When the window is closed at night, the cold frame retains the heat that the soil absorbed during the day.

To receive maximum winter sunlight, orient the cold frame toward the south and away from prevailing winds. Soil drainage in the cold frame must be excellent. Painting the inside walls white or silver helps reflect more light to the plants.

Many vegetable gardeners use a cold frame to start seeds early. A cold frame lets you start most garden seeds up to eight weeks earlier than in the garden. In short-season areas, those eight weeks really count.

If your winter is not too severe, you can grow winter salad greens easily in a cold frame. Lettuce, chives, and most other greens will continue growing until spring if protected in this way. The heat stored in the soil will carry the greens through the winter.

Cold frames can also be used for forcing bulbs and propagating a variety of flowers, azaleas, and trees and shrubs.

Hotbeds are similar to cold frames but have a heat source inside, usually an underground cable, to provide heat to the soil and plants. At one time, the heat source was a thick layer of manure tightly packed under the soil.

## PLANT SUPPORTS

Many vegetables need help to stay upright, make the most efficient use of space, and produce a better crop by keeping the fruit off the ground, where it will be subject to damage from insects, diseases, and wet soil. Climbing plants also receive more sun and bear earlier if trained upright. Although climbers can be grown on the ground, they produce better if they are not. Climbing vegetables include pole beans, tomatoes, cucumbers, squash, melons, and peas.

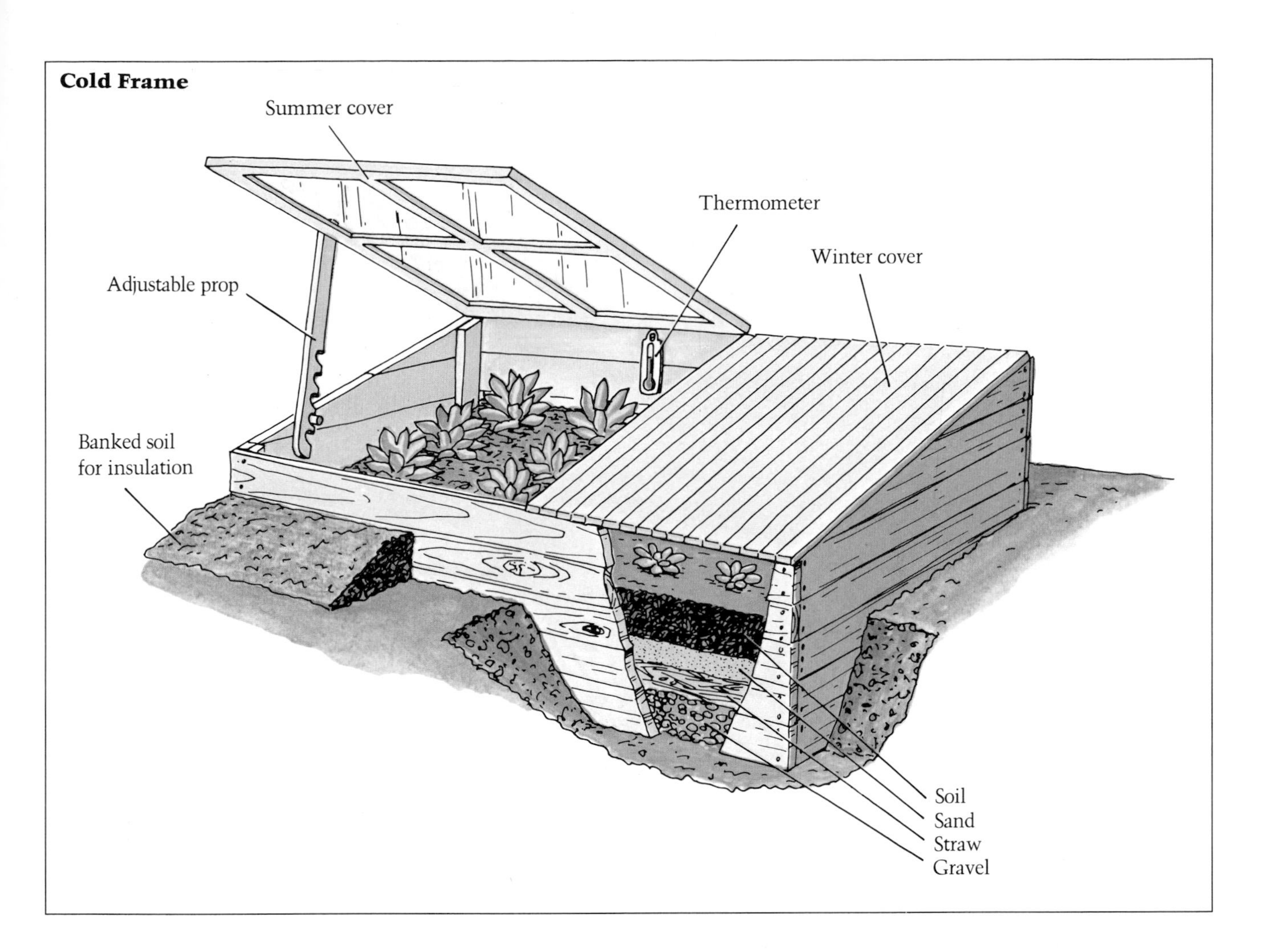

*Opposite, top: The top half of a large plastic jug protects young plants from sun, wind, and garden pests. Opposite, bottom: This simple cold frame consists of a brick perimeter topped with a sheet of glass. Right: A thermocouple causes this commercial automatic cold frame to open when the air temperature exceeds 72° F and close when it drops below 68° F.*

Tomatoes can be staked to wooden or bamboo poles, grown on string, or grown in cages. You can buy cages or make your own from wire mesh; just make sure to buy mesh large enough that you can get your hand through it to harvest the crop. If you have selected a tomato variety that grows large and bears heavily, make your cage strong enough to support it.

Other vegetables can be staked with PVC pipe, bamboo, thin branches, poles, 2 by 2s, or any similar support. Poles and other supports can be constructed in a flat plane or in the shape of a tepee or an A-frame. Vegetables can also be trained along a fence at the perimeter of the garden. Trellises can be strung with fishing line or string to help the vines grow. Wire and metal stakes can be used, but they are the least desirable, because they get very hot in the sun and can burn tender new growth and tendrils.

## WINTER PROTECTION

Most vegetables are planted and harvested in one growing season, leaving the ground bare of a crop over the winter. However, a few vegetables, such as asparagus and rhubarb, are true perennials, and some root crops, such as parsnips, can be left in the ground over the winter. When temperatures fall so low as to damage or kill these plants, winter protection is needed.

A winter mulch will keep the ground warmer and prevent the alternate freezing and thawing of the soil, which is the main culprit of the cold. Many mulching materials can be used: garden soil brought in from another area, chopped oak leaves, salt hay, straw, even shredded newspapers. Apply mulch after the first frost but before the ground freezes, and remove it gradually in spring as the first signs of growth appear.

## PESTS AND DISEASES

In general, it's best to select vegetables and varieties that aren't susceptible to the insects and diseases prevalent in your area. Use barriers, such as row covers and film mulches, to keep insects away from the plants.

Here are some ways to recognize and control the most common pests. To find out about additional insects and diseases, see *The Ortho Problem Solver,* available for your use in most garden centers. If you have difficulty solving your garden problem, call the county agricultural extension agent.

When using insecticides on vegetables, read the label carefully and follow the directions exactly. The label will tell you which plants you can safely spray and how long to wait before harvesting.

*Left: Supporting plants that would otherwise sprawl can save space, improve yield, and make harvesting a little easier.*
*Right: When covered with clear plastic film, a wire cage protects young tomato plants from the wind.*

### Aphids

Although aphids feed on many vegetables, they are most damaging to members of the cabbage family, which includes broccoli, cauliflower, brussels sprouts, and cabbage. Growth in damaged plants is stunted and often distorted. Aphids are also responsible for spreading many diseases. Different species are different colors including gray, black, brown, and red.

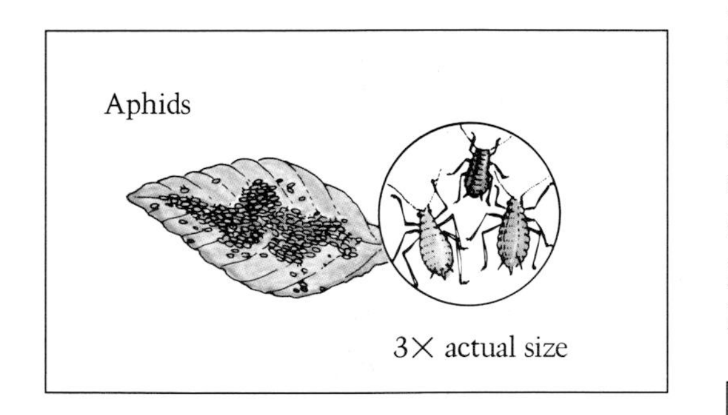


**Control** First try simply washing off the aphids with a water spray. This is often effective if repeated frequently enough. If that isn't sufficiently discouraging, spray them with a soap-based insecticide or one containing pyrethrin, diazinon, or malathion.

### Beetles

This incredibly diverse group of insects includes harmless, damaging, and beneficial types. Ladybugs are among the beneficial variety, since they eat garden pests; the cucumber beetle (*Diabrotica*) is one of the most troublesome. Other pestiferous beetles are illustrated.

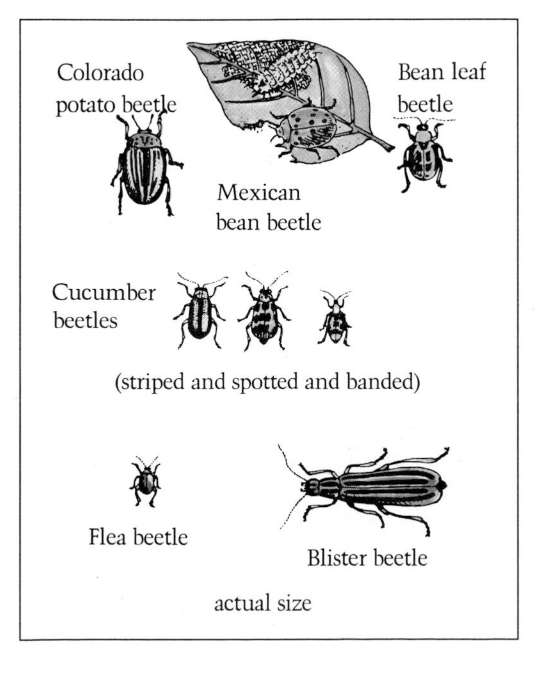


**Control** Beetles can often be kept away from plants with row covers. Many chemical controls are available, but take care to use only those that are made specifically for vegetable gardens. Products containing diazinon, rotenone, Sevin®, and methoxychlor control many beetles. Small numbers of beetles can be picked off plants by hand.

### Bugs

Nongardeners may consider any insect a bug, but when gardeners talk about bugs, they mean a specific group of insects, many of which are significant garden pests. Stinkbugs, for example, attack carrots, lettuce, okra, and peppers. Squash bugs damage squash and pumpkins.

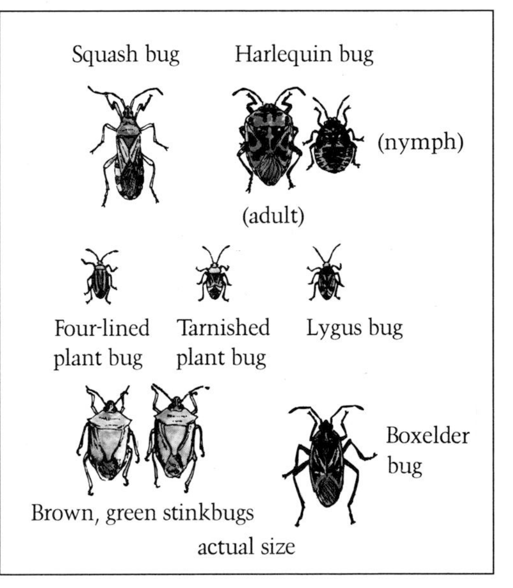


**Control** The best control is with products containing Sevin®, methoxychlor, pyrethrin, or malathion. Apply while the bugs are still young; mature bugs are harder to kill.

### Caterpillars and Worms

These are the larvae of various moths and butterflies. They come in all sizes and colors, some hairy and some with spines. All have healthy

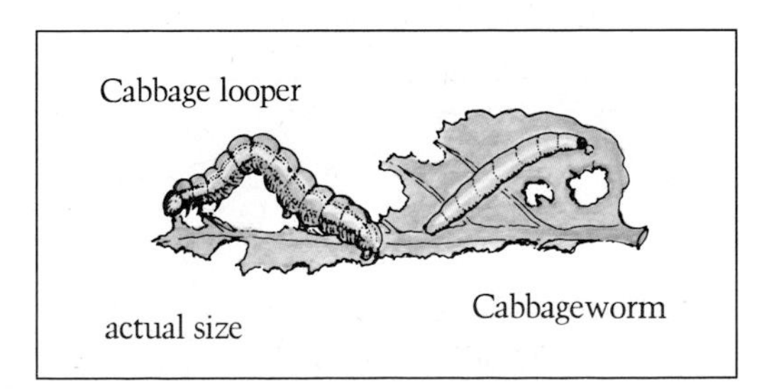

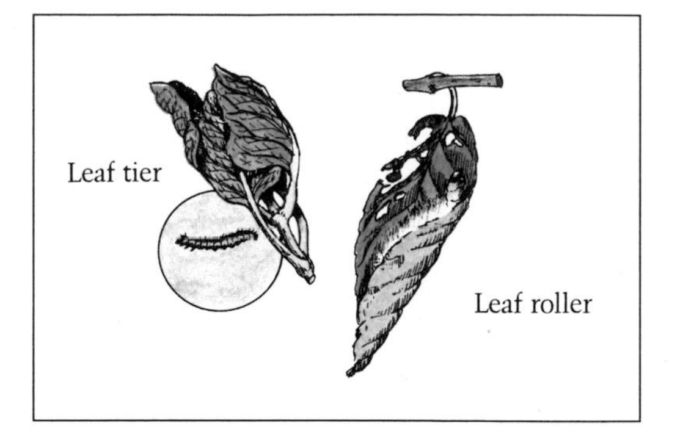


appetites and are among the most frequently damaging pests in vegetable gardens. The cabbage looper and tomato hornworm are some of the most common caterpillar pests.

**Control** Handpicking works for small numbers of caterpillars, and for large individuals such as the tomato hornworm. Or spray with the biological control Bt (*Bacillus thuringiensis*), which is a bacterial disease of caterpillars. Other controls include Sevin®, diazinon, pyrethrin, rotenone, or malathion.

### Corn Earworms and Tomato Fruitworms

Both these names refer to the same insect; which name is used depends on which crop is attacked. The corn earworm causes more damage to sweet corn in the United States than does any other insect. The worst damage occurs during the silk stage, when the eggs hatch on the silk and the larvae begin eating their way into the developing ear, destroying the kernels and leaves. The tomato fruitworm attacks the flowers and developing fruit of tomatoes.

**Control** Since corn earworms overwinter as pupae in the soil, thorough soil cultivation in late fall will aid in control. Look for sweet corn varieties with some resistance to this pest. Feeding worms can be stopped with insecticides that contain Sevin®. Treat the tomatoes with Sevin® when the fruit is about ½ inch in diameter.

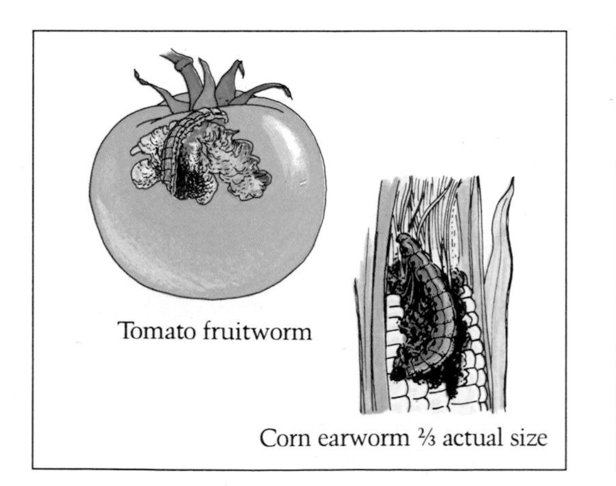


### Cutworms

The cutworm, a type of caterpillar, is extremely prevalent in the garden. During the day it lives just beneath the soil surface; at night it emerges to feed on the tender stems of young tomatoes, cabbage, peppers, beans, and corn, cutting them off at the soil line.

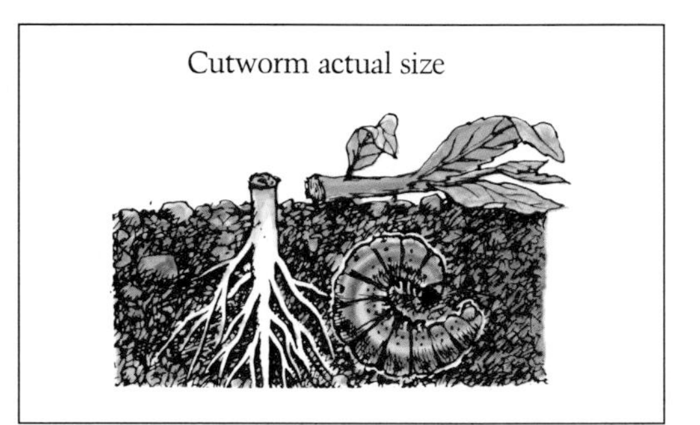


**Control** Thorough cultivation in late summer or early fall will expose and destroy many cutworm eggs and pupae. You can also make a collar out of an empty tin can or a cylinder of builders paper and insert it about an inch into the soil to protect the young plants. Cutworms won't climb over the collar. Or use a granular product containing diazinon or chloropyrifos before planting.

### Leafhoppers

These small (⅛ to ½ inch), wedge-shaped insects feed by piercing and sucking. Although they are general feeders, they particularly damage beans, lettuce, potatoes, squash, and tomatoes. They will also feed on bean blossoms, causing a poor pod set. Often the most severe leafhopper damage is caused not by their feeding but by the virus diseases they spread.

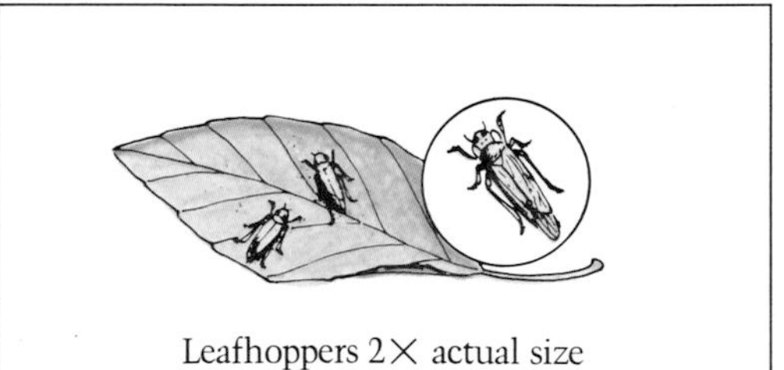
Leafhoppers 2× actual size

**Control** Reflective mulches, such as aluminum foil, are helpful, especially when plants are young. Sprays containing insecticidal soap, malathion, pyrethrin, rotenone, or Sevin® also are effective.

### Leaf Miners

These troublesome insect larvae live inside plant leaves and mine the plant tissue, frequently rendering the leaves useless. This reduces the plant's vigor and, of course, the harvest. Peppers, tomatoes, cucumbers, melons, and squash are often attacked by these pests.

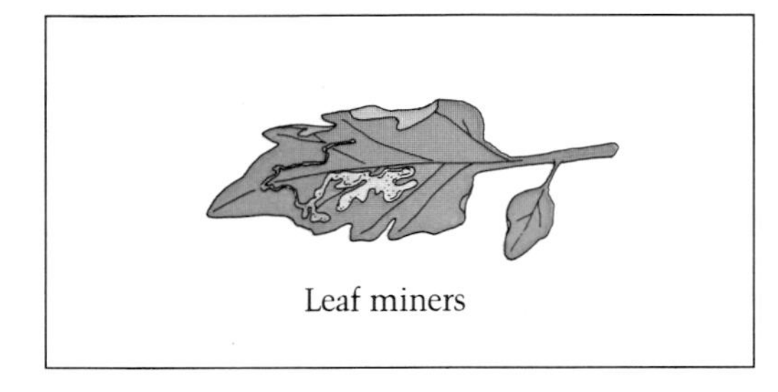

Leaf miners

**Control** Use a spray containing malathion or diazinon, according to label directions.

### Mites

The webbing that these tiny spiders deposit on the undersides of leaves is often more visible than the insects themselves. Thriving in hot, dry, and dusty conditions, mites make a speckling on the top of damaged leaves. This always reduces the plant's vitality; sometimes it destroys the plant altogether.

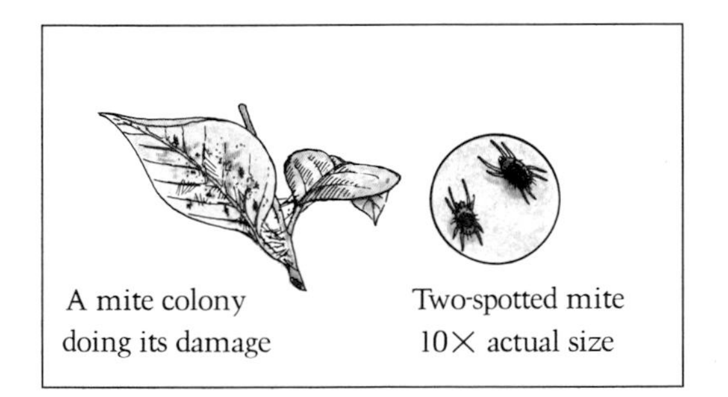

A mite colony doing its damage

Two-spotted mite 10× actual size

**Control** Use a malathion spray on beans, peas, broccoli, and brussels sprouts. If melons and squash are attacked, use a diazinon spray or treat with insecticidal soap.

### Onion and Radish Maggots

These are the larvae of flies that appear in spring and lay eggs on the soil near the base of vegetables. Onions and radishes are the two favorite targets, although other vegetables are attacked as well.

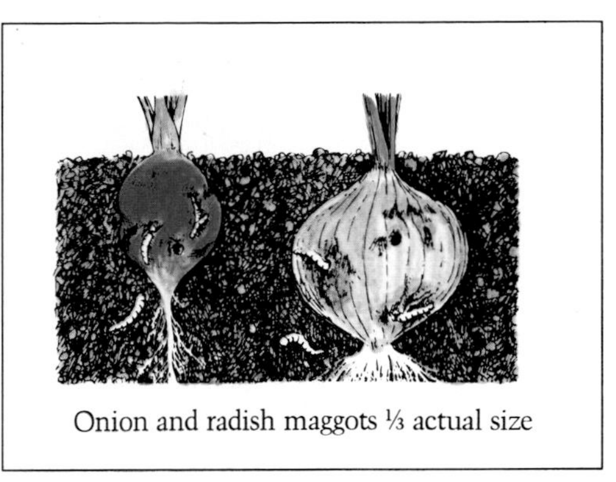

Onion and radish maggots ⅓ actual size

**Control** A fine-mesh wire cloth over the seedling row will prevent the adult flies from laying eggs. Or use a granular or dust product containing diazinon or chloropyrifos. Control for the cabbage root maggot is similar.

### Slugs and Snails

These are among the most ubiquitous and damaging vegetable-garden pests. They feed mostly at night, hiding in cool, damp locations during the day. The silvery slime trail they leave behind is the mark of their activity.

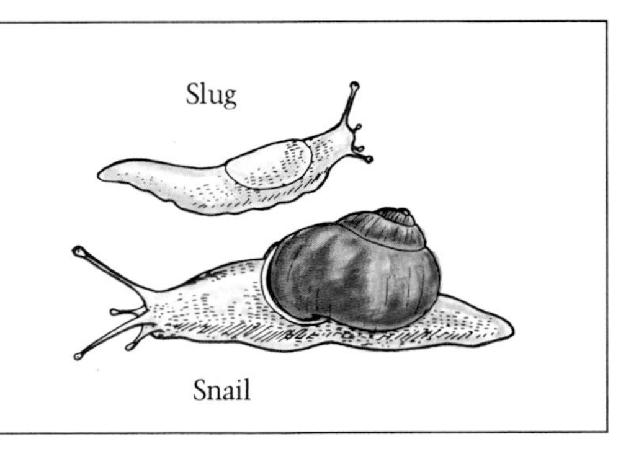


**Control** If you can, tack a copper-plated strip to a board all around the garden or around each bed. Slugs and snails will not cross it. Thick liquid baits containing metaldehyde can be used as a barrier around the garden. Renew it every couple of weeks. Or scatter bait pellets containing metaldehyde all through the garden area every couple of weeks.

### Whiteflies

Adult whiteflies are small ($^1/_{16}$ inch long), wedge shaped, and pure white. They fly like clouds of snowflakes when disturbed. Nymphs, immature whiteflies that are even smaller than the adults, do the most damage. They are scalelike, flat, and oval, and can be pale green, brown, or black. They stunt plant growth by sucking the juices from the leaf undersides.

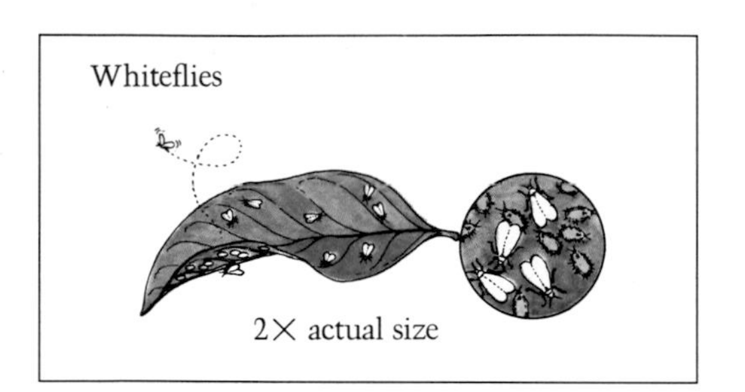


**Control** Use diazinon, pyrethrin, or malathion products intended for use on food crops. Follow label directions and be sure to spray the leaf undersides, where most whiteflies hide. Since yellow attracts whiteflies, yellow sticky traps similar to flypaper are effective. Insecticidal soap also works well.

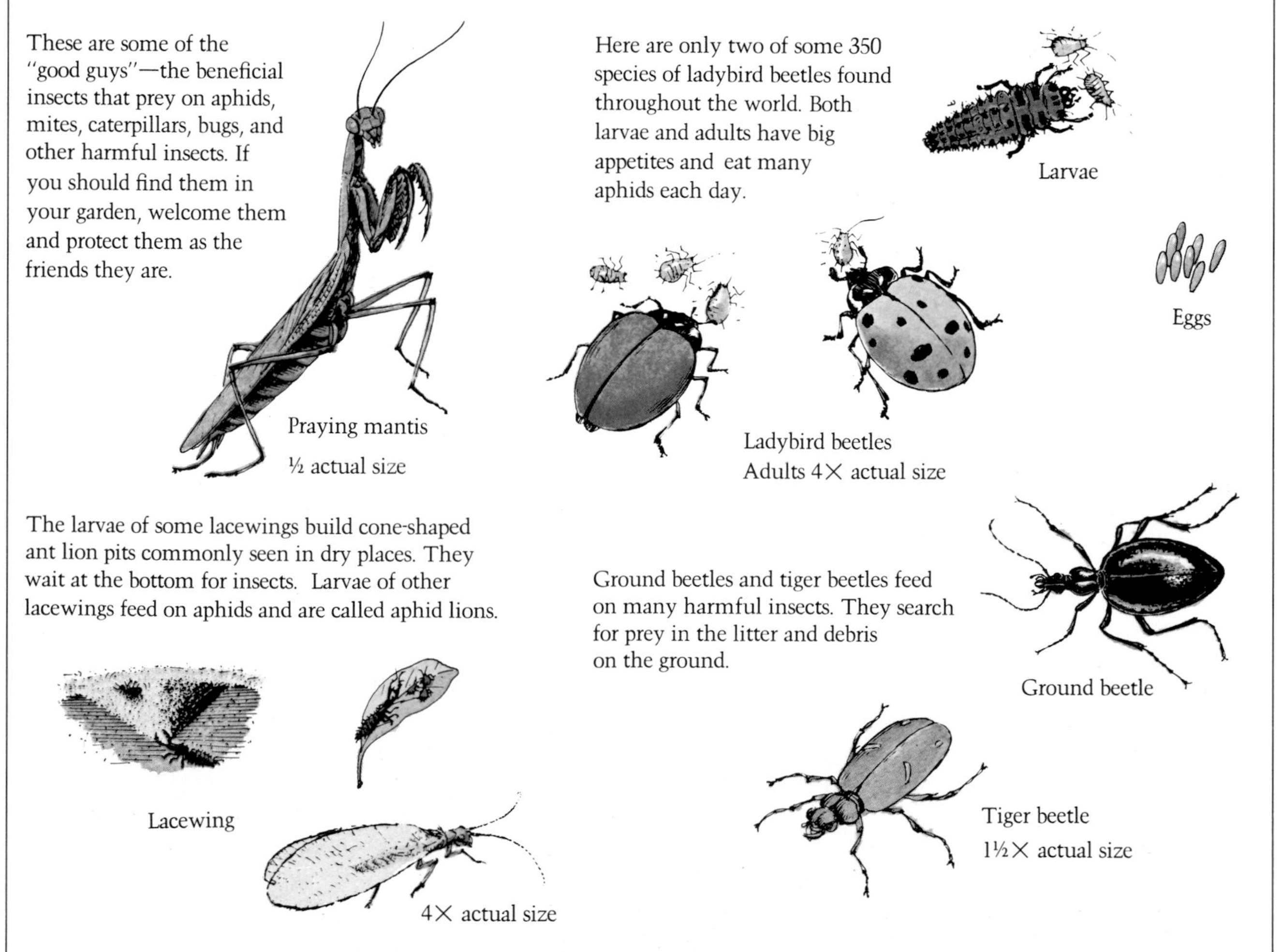

### Using Plants to Control Insects

The idea of using certain plants to repel insects is an old one, and many gardeners conscientiously observe certain planting patterns; however, the results are mixed. If you wish to try them, here are some popular repellent plants. Garlic is said to keep aphids away. Radishes, when planted near beans, cucumbers, eggplant, squash, and tomatoes, are said to discourage mites and beetles. Nasturtiums are supposed to rid the garden of most aphids, beetles, and squash bugs. Mexican bean beetles are discouraged by planting potatoes near the beans. Marigolds repel nematodes and the Colorado potato beetle. Asparagus beetles are kept away if tomatoes are planted near the asparagus.

### Preventing Diseases

Planting disease-resistant varieties is the surest way to prevent diseases from devastating your crop. Since more disease-resistant varieties are bred each year, you have a wide choice.

If you have already had—or want to avoid—trouble with diseases, check the variety names of individual vegetables in the chapter starting on page 69 (we have noted those that have some disease resistance) and in seed catalogs. In some cases there is no control for the disease, and the only cure is not to get it in the first place. Be especially alert for mention of resistance to the diseases that damage these vegetables.

- ☐ Cabbage: virus yellows
- ☐ Cucumbers: scab, mosaic, downy mildew, powdery mildew, anthracnose
- ☐ Muskmelon: fusarium wilt, powdery mildew
- ☐ Snap beans: mosaic, powdery mildew, root rot
- ☐ Spinach: blight, blue mold, downy mildew, mosaic
- ☐ Tomatoes: fusarium wilt, verticillium wilt, tobacco mosaic virus, alternaria

Many fungus diseases are spread by splashing water and are favored by wet conditions. Avoid watering in the afternoon or at night so that the leaves will not be wet for extended periods, which is when fungal spores germinate. Avoid crowding, which leads to poor air circulation, if diseases are a problem. Poorly drained soil also contributes to fungus diseases. Many viral diseases are spread by aphids, which can easily be controlled. Keeping the garden weed free also reduces the chance of disease.

Controlling diseases by rotating crops is good, standard gardening advice, because many diseases are soilborne. The most effective method is to change the garden site every few years. Second to that is rotating the crops within the garden so that the same crop or a crop from the same family doesn't occupy the same space year after year. However, it's difficult to rotate crops in a 10- by 20-foot garden.

*Opposite: Inspect plants regularly for insects or for signs of damage. Early diagnosis allows you to treat problems more effectively.*
*Bottom: When using overhead sprinklers, water early in the day so that plant leaves will dry quickly.*

## WEEDS

Unfortunately, weeds compete with vegetables, thriving in the sun and rich, moist soil the vegetables need. Every time you work the soil, you inadvertently bring weed seeds closer to the surface, where they have a better chance to germinate.

Weeds should be pulled from the garden as soon as they appear. They compete with vegetables for sun, water, and nutrients; decrease air circulation, which leads to disease; and are breeding sites for many insects.

Weeds can be pulled by hand or with a hoe. Be careful when weeding not to disturb and damage the vegetables' roots. Weeding after rain or watering makes it easier to remove the undesirables from the ground.

A good weed preventive is mulch. Several types are available. Black plastic is widely used; punch holes in the plastic so water can pass through. Gardeners who don't like the look of plastic in the garden can cover it with an organic mulch. Organic mulches can also be used by themselves. See page 17 for more information on mulches.

Preemergent herbicides such as DCPA prevent weed seeds from germinating. Read the label carefully and follow directions; determine which plants they can be used around safely.

Chemical weed killers should never be used to kill existing weeds in the vegetable garden for fear of damaging the crop. Glyphosate can be used, however, on a plot of land that is not yet in use to destroy weeds before soil preparation and later planting. Read label precautions as to the length of time you must wait between applying glyphosate and planting.

Many beginning gardeners complain that the seed packets they purchased contained weed seed. This is simply not true. With today's highly developed techniques of cleaning seed, the chances of weed seed appearing with crop seed are negligible. There is a thousand times more weed seed in the soil than there is in a packet of crop seed.

*When hoeing weeds, slice or scrape the weed just below ground level.*

## ANIMALS AND BIRDS

Gardeners, especially those in suburban and rural areas, compete with birds and animals that eat seeds, seedlings, and crops. Plastic or wire netting or mesh can be laid over the vegetables when needed, and rolled up and stored until needed again. Netting can also be stretched over wooden frames and placed over the beds as needed; make sure it is fine enough to keep out mice and small birds.

Knee- to waist-high woven fencing can keep out poultry, rabbits, and dogs and also double as a trellis. Higher fences will be needed for deer, which are also deterred by repellents, sprays containing the fungicide thiram, and hot pepper flakes. Dried blood repels some animals but has to be reapplied after watering and rain. Remember that dried blood is a source of nitrogen, so recalculate your fertilizer needs if using it as a repellent.

Moles and rodents are best controlled with bait, traps, or a repellent. Planting in raised beds and lining the bottom of the beds with fine chicken wire will also keep out many of these ground-dwelling pests.

*Top: A wire cage protects a single row of lettuce from birds, rabbits, and other animal pests. Bottom: A wood frame covered with wire mesh safeguards an entire bed of strawberries from pests.*

CELERY
Utah

# Planning the Vegetable Garden

*A major key to the success of your garden is planning. Information on succession plantings and interplanting will help you estimate the harvest and make the best use of the garden space. Knowing the needs of your family will help you to decide what and how much to plant; selecting the right varieties for your area and taste will help ensure success.*

Vegetable gardens bring many joys, but they also bring problems. One of the most typical (and most frustrating) occurs when everything ripens at once—a classic illustration of the perils of too much too soon. It isn't easy to plan for a continuous harvest of fresh vegetables, especially for the beginner. And the limited-space gardener has a thornier challenge than the farm-space gardener.

If you're fortunate enough to have plenty of room, you can block out space for the spring garden and leave some space empty in readiness for the summer garden. This way, you can plan and plant the summer garden without interference from the spring garden.

But a city gardener with a 20- by 30-foot area needs to plan for both the spring and summer gardens (and also for the fall and winter gardens in long-season areas). Clutching a dozen packets of seed, each enough to plant 50- or 100-foot rows, the small-space gardener waxes optimistic and pictures row upon row of beautiful, healthy plants.

But although it might seem as though 20 or 30 heads of lettuce couldn't possibly produce enough salads for the whole family, when the 30 heads all mature within a 10-day period, you know you've

*Growing vegetables will be more rewarding if you plan carefully: Map out your space, choose suitable varieties, figure the planting dates, and prepare for harvests.*

Top: *A well-organized vegetable garden is a joy to tend.*
Bottom: *Flowering causes onion bulbs to shrink and toughen; pull up the onions right away.*

planted more than enough. A dozen cabbage plants don't seem excessive in their little nursery trays. But they'll make 30 to 40 pounds of cabbage when they are mature. That's a lot of sauerkraut.

Garden planning—making the most efficient use of the garden space you have—is the best way to avoid such situations. It's always wise to plan the garden on paper before you pick up a trowel for the first time. That way you'll know what to plant, how much to plant, and how to time the harvest so that you can make the best use of the available space.

## LENGTH OF HARVEST PERIOD

The first considerations in selecting vegetables are the time it takes them to reach maturity and the length of the harvest season. Some vegetables mature quickly, leaving their space free for future plantings. Others have long harvest periods and occupy the garden space all season. To avoid having all the corn or cabbage ready at the same time, select varieties with a different number of days to maturity to spread the harvest out over a longer period of time. Selecting different varieties also lets you try out new types.

The following vegetables have long harvest periods.

- ☐ Green beans begin producing after 50 to 60 days and continue until frost.
- ☐ Eggplant begins in 60 to 70 days and lasts until frost.
- ☐ Peppers (both bell peppers and chiles) begin after 60 to 80 days and last until frost.
- ☐ Summer squash is harvestable after 50 to 60 days and keeps going until frost.
- ☐ Tomatoes begin anywhere from 50 to 90 days after planting and last until frost.

In contrast, corn and radishes are harvested within a two-week period. The shorter the duration of the harvest, the more important small, successive plantings are.

Some root crops, including carrots, beets, parsnips, and salsify, will last a long time in the soil. Of these, carrots and beets provide a succession of harvests—you can begin harvesting when they are baby sized and continue to harvest up to maturity—but they will also last in the soil for periods of weeks or months, depending upon the climate and time of year.

Since leaf vegetables such as leaf lettuce and Swiss chard can be picked a leaf at a time, as needed, they have long harvest periods. Depending upon the time of year, this period may last several weeks or more.

Perennial vegetables—asparagus, rhubarb, and Jerusalem artichokes—come back year after year.

## SUCCESSION PLANTINGS

One way to ensure a continuous harvest of vegetables that mature quickly is to make successive plantings of small quantities. For example, if you first plant lettuce in March, you can make another planting in April and another in May. Successive plantings of snap beans, four to six weeks apart, will give you fresh beans for five months or more. To get a continuous harvest of a vegetable, you must plant the second planting before you harvest the first.

When you chart a succession of plantings on paper, the goal of a long harvest season comes within reach. It's easier than you might think. Planting times are determined by four factors: when the soil is workable in spring; the date of the last spring frost; the date of the first fall frost; and the average date that the soil becomes frozen, if ever. Next, look up the vegetables you would like to grow, and note when they should be planted and the average number of days until harvest. By taking all these figures into consideration, you'll know what to plant when, about when it will be ready to harvest, and when to replant for a steady harvest. Once you've charted plantings in this way, you can make adjustments easily that will lengthen the harvest season.

## INTERCROPPING

Also known as double cropping, intercropping is another way to plan the garden to increase yield. The principle of intercropping is to plant fast-maturing crops with those that have a long time to maturity. When the first crop is ready, it is harvested, leaving space for the later-maturing vegetables to grow.

### *A Typical Garden Calendar*

You can plan the succession planting and harvest in your garden by making a calendar like this of your area. This one is for the Delaware area, taken from the University of Delaware Extension Service Bulletin 55. Adjust your chart based on your climate and planting dates.

| Mar | Apr | May | June | July | Aug | Sep | Oct | Nov |
|---|---|---|---|---|---|---|---|---|
| | | Snap beans | | | Snap beans | | | |
| | | | Lima beans | | | Lima beans | | |
| Cabbage | | | Cabbage | | | | | |
| | | | | Cabbage | | | Cabbage | |
| | | Cucumbers | | | Cucumbers | | | |
| | Carrots | | Carrots | | | | | |
| | | | | Carrots | | | Carrots | |
| | Beets | | Beets | | | | | |
| | | | | Beets | | | Beets | |
| | | | | Broccoli | | | Broccoli | |
| | | | | | Cauliflower | | Cauliflower | |
| | | Cantaloupe | | | Cantaloupe | | | |
| Lettuce | | | Lettuce | | Lettuce | | Lettuce | |
| | Onions | | Green onions | | Onions | | | |
| Peas | | | Peas | | | | | |
| | | | Peppers | | | Peppers | | |
| Radishes | | Radishes | | | | Radishes | Radishes | |
| Spinach | | Spinach | | | | Spinach | | Spinach |
| | | Sweet corn | | | Sweet corn | | | |
| | | Squash | Squash | | | | | |
| | | | Winter squash | | | | Winter squash | |
| | | Tomatoes | | | | Tomatoes | | |
| | Turnips | | Turnips | | | Turnips | Turnips | |
| | | Watermelon | | | Watermelon | | | |

Key: Plant Harvest

**Planting Seasons**

**Early Spring**

Plant as soon as ground can be worked in spring: broccoli plants, cabbage plants, endive, kohlrabi, lettuce, onion sets, parsley, peas, radishes, spinach, turnips.

**Midspring**

Plant these at time of the average last killing frost: beets, carrots, cauliflower plants, onion seeds, parsnips, Swiss chard. Plant two weeks later: beans, corn, early tomato seeds, potatoes.

**Early Summer**

Plant when soil and weather are warm: cantaloupe, celery plants, Crenshaw melons, cucumbers, eggplant plants, lima beans, pepper plants, potatoes for winter, pumpkins, squash, tomato plants, watermelons.

**Midsummer-Fall**

Plant in late June or early July: beets, broccoli, cabbage, cauliflower, kohlrabi, lettuce, radishes, spinach, turnips.

Radishes can be sown with carrots. The radishes mature quickly and are pulled at about the time that the carrots need additional space. Onions can be interplanted with almost anything and used for scallions when the companion crop needs more room. Bean seeds can be sown between lettuce plants that are halfway to maturity. When the lettuce is ready to be harvested, the beans can take over. There are many combinations; choose almost any small, fast-growing vegetable and plant it between larger ones that take a long time to mature.

## COORDINATING WITH THE KITCHEN

If your family doesn't like spinach, there's little point in growing it. But if your preferences are for a salad garden, certainly include spinach along with lettuce, chard, onions, cress, chicory, radicchio, and tomatoes. Make a list of your family's favorite vegetables before deciding what to plant, and then determine how much space you have and how many plants will fit into it.

The National Garden Bureau (an educational service of the seed industry) suggests the following major considerations to keep in mind when deciding what to plant where.

- ☐ Do you and your family like the vegetable?
- ☐ How many days are required from planting to harvest?
- ☐ Does the vegetable prefer cool or warm weather?
- ☐ How large do the plants grow?
- ☐ How many plants of each kind will you need to feed your family?

## A SAMPLE GARDEN PLAN

Illustrated on the next page is a plan for a 25- by 30-foot garden area that will bring a succession of crops throughout the growing season. A small-space gardener can either use this scheme as is or adapt it by using pieces of it; gardeners with more room can increase the number of plants or the types of vegetables grown.

Divide the area into three sections, each approximately 10 feet wide. Plant the first block in spring, as soon as the soil is workable, with early, cool-weather crops, such as carrots, beets, spinach, onions, lettuce, radishes, turnips, early cabbage, and early potatoes. If you allow for proper spacing between rows (see the chart beginning on page 63), this planting will use a 10-foot depth of the first block.

In the second 10- by 10-foot block in the next section, plant the same vegetables but a month later. Or you could try other vegetables—for example, parsnips, chard, or endive in place of the spinach, onions, and radishes.

Plant the third 10- by 10-foot block in the last section with those crops that will mature in fall. Turnips, lettuce, spinach, radishes, chard, kale, mustard, and Chinese cabbage are some cool-weather crops that will benefit from being planted after the hottest days of summer have passed.

After the soil is thoroughly warm, plant the rest of the garden space with warm-season crops. These crops include cucumbers, melons, summer squash, tomatoes, corn, pole beans, bell peppers, and chiles.

Plant the tomatoes in 18-inch-high wire cylinders. Train pole beans along a wood-and-string trellis. Plant the corn in three blocks of three rows, making three plantings spaced two weeks apart.

## LAYING OUT THE GARDEN

Once you have a garden plan on paper, it needs to be interpreted outdoors. Vegetable gardens are best oriented north to south, with the taller vegetables at the north end so they do not shade the lower-growing ones. A north to south orientation also gives all of the plants an equal share of the noonday sun.

If you're laying out a garden for the first time, measure the area carefully and mark the boundary with strings so it will be straight. Also measure the individual rows or blocks before planting and mark them with string as well so the space is delegated as planned. There are several planting systems you can choose from.

Garden Plan

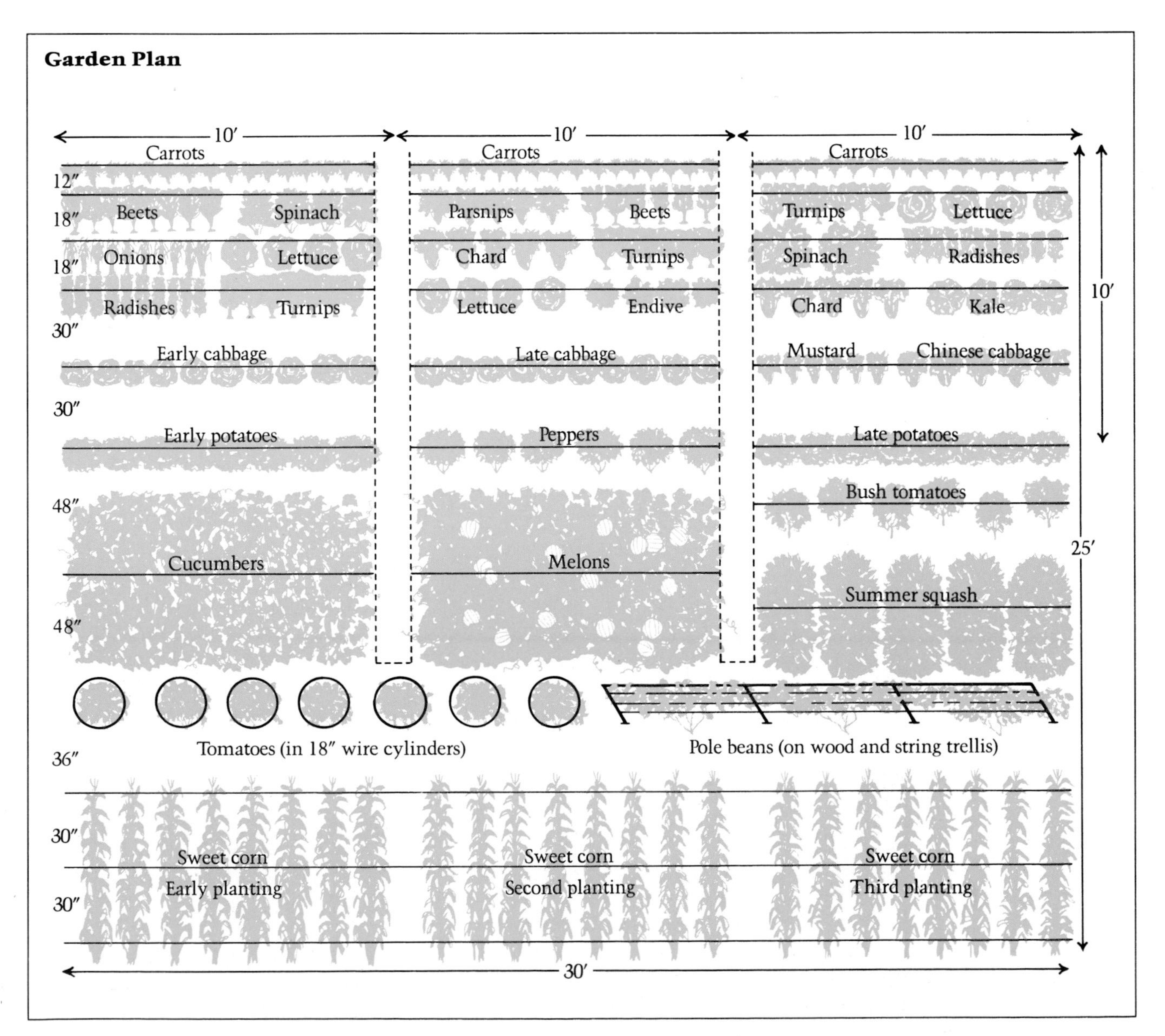

### Single Row

The most traditional system, the single-row method uses vegetables in a single row with a path between rows wide enough for walking and carrying gardening tools. It takes up the most space, and the space between rows needs to be kept clear of weeds, but with its wide paths and narrow rows, it is easy to work in.

### Wide Row

One of the best ways to increase yield without increasing the size of the garden plot is with wide-row planting. Individual plants may not be as productive as with the single-row system, but the yield will be greater for the area planted. Rows are 3 to 5 feet wide; the number of plants across the row depends on the type. Paths are left between the rows as with the single-row system. Wide rows eliminate a lot of weeding because the plants grown close together choke out the weeds. The heavy foliage cover also keeps the ground cooler and can extend the season for cool-season vegetables, such as lettuce.

### Raised Beds

Raised beds increase yield over both single-row and wide-row methods. Beds are raised 4 to 6 inches off the ground and the soil within them is improved. They can be up to 5 feet across; any wider and they will be difficult to work without stepping into them. The soil in raised beds dries out faster in the spring, allowing earlier planting. It also warms up quicker. The beds can be simple mounds of soil or can have walls of wood or stone. Raised beds are often permanent, so paths can be improved.

### Intensive Gardening

This system is an intensification of raised-bed gardening. Individual planting boxes or beds are constructed with 1 by 8s or similar boards and the area is filled with improved soil. Sometimes the entire garden is double-dug. Each unit is planted to capacity with one crop. As the cool-season crops are harvested, they are replaced with warm-season vegetables. In fall, they are again replanted with cool-season vegetables.

*Planting in wide rows, or blocks, instead of single rows makes more efficient use of space and discourages weeds by shading all the soil.*

## THE GARDENER WITHOUT A LARGE GARDEN

Most gardeners would like to have a large, country-style garden, but most don't. Nevertheless, small-space gardeners are a determined breed. Productive vegetable gardens have been grown in small backyards and on rooftops, balconies, decks, paved driveways, and other nontraditional locations.

Of necessity, the gardener without a large vegetable garden can neither overplant nor waste the harvest.

Some determined gardeners combine resources with their friends and neighbors to make a community garden. Such an undertaking is simultaneously practical, social, and educational—practical because it makes good use of the vacant lots found in urban areas and lets the gardeners grow food close to where it is used, social because it offers a great way to get to know your neighbors, and educational because of the gardening knowledge that's shared.

### Vegetables in Containers

Versatility is the hallmark of the container garden. On balconies, decks, or patios, vegetables can be grown in boxes, tubs, bushel baskets, cans, and planters of all shapes and sizes. Any of these containers will work; the depth is what's critical.

The major considerations to keep in mind are portability and frequency of watering and fertilizing. The shallower the container the more frequent the need for water and fertilizer. Depending on the air temperature and the size of the container, daily watering might be needed. For the following vegetables, the

*Top: This mounded bed is intensively planted with mixed salad greens. Bottom: Lettuce thrives in a vertical planter.*

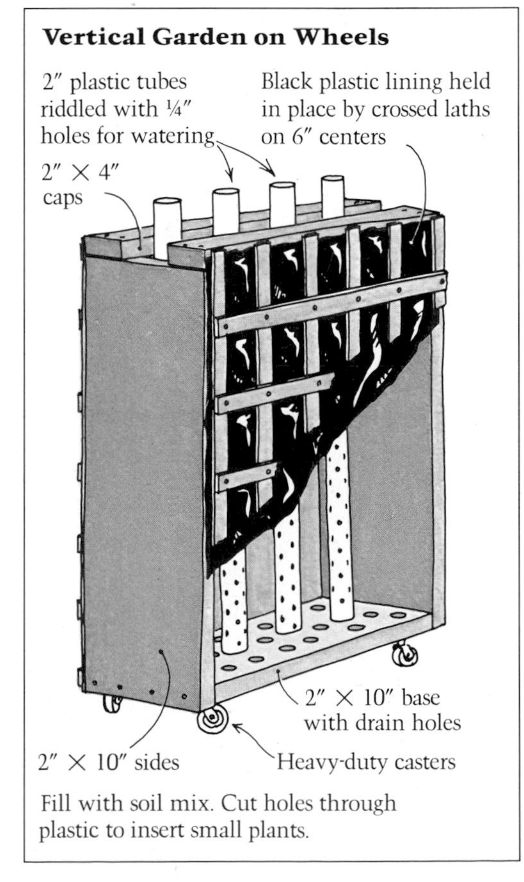

*Limited space or poor soil does not daunt the determined vegetable gardener. Enough vegetables to feed a family of four throughout the growing season is produced in this array of containers.*

containers should have these minimum depths.

- ☐ 4 inches deep—lettuce, turnips, radishes, and beets.
- ☐ 6 inches deep—chard, kohlrabi, short carrots.
- ☐ 8 inches deep—bush beans, cabbage, chiles, eggplant, bell peppers, and bush cucumbers.
- ☐ 10 inches deep—cauliflower, broccoli, and brussels sprouts.
- ☐ 12 inches deep—parsnips, salsify, long carrots, and tomatoes.

The choice of which vegetables to plant in container gardens also depends on which ones give the highest return per square foot of space—in other words, those that can be spaced most closely in the row. (See the planting chart on pages 63 to 65.) Some vegetables in this group are carrots, beets, chives, leaf lettuce, mustard, green onions, radishes, and turnips.

For a continuous harvest, you might plant a total of six containers—two early, two more three weeks later, and the last two containers two weeks after that.

How vegetables grow in your climate will dictate your choice of late spring and fall plantings. Container plantings make it easy to think of harvests in terms of the number of meals instead of the total quantity of plants.

### Small-Space Varieties

Gardens that cannot grow out can grow up. A fence 5 feet high and 20 feet long provides a whopping 100 square feet of growing space. It can be covered with vines of pole beans, cucumbers, tomatoes, squash, and gourds without infringing on most of the ground space.

The needs of small-space gardeners have not gone unheeded by vegetable plant breeders; they have come up with many miniature and bush varieties of popular vegetables. If you haven't the room for vining cucumbers, try 'Pot Luck', 'Bush Champion', 'Bush Crop', or 'Salad Bush'. The full-sized cukes they produce are almost as big as the entire plant. Also consider the bush-type squashes, such as 'Table King' and 'Butterbush', or the new space-saving pumpkin, 'Spirit Hybrid'. 'Honeybush' muskmelon is as luscious as a melon from a full-sized plant. There are many tomatoes that can be grown in patio containers or hanging baskets.

## RECORD KEEPING

As good as people think their memories are, they do tend to forget, so record keeping is important to ensure success in following years. A simple record book of what you planted, how long the seeds took to germinate, how long the crop needed to mature, whether there was too much or too little to harvest, and notes about particular varieties that did (or didn't do) especially well is invaluable in planning the garden for future years.

Labeling is also critical. Whether you start your own plants from seed or buy transplants, label all of them with the variety name and planting date so you have this information for future reference. White plastic labels that can be written on with a pencil or waterproof ink are best, since the weather won't wash away the notations. A map of the garden is also a good idea in case the labels are lost.

### *All-America Selections*

You may have run across the designation All-America Selection. Catalogs frequently cite these honored plants, and many are noted in this book. Just how a plant wins the title and what it means are of importance to the gardener.

The All-America Selection organization includes a council of judges and some 26 vegetable test gardens located in different climates throughout the United States and Canada. Its purpose is to evaluate the new vegetable (and flower) varieties introduced each year. The entries, known to the judges by number only, are grown in trial rows side by side with the most similar varieties already available. Judges vote primarily on home garden merit, paying special attention to climatic adaptation and vigor.

Of the 75 to 100 entries each year, usually only 3 or 4 earn enough points to be given the All-America Selection medals. The gardener thus can be assured that an All-America Selection is superior in its class and is adapted to a wide climate range.

## CHOOSING VARIETIES

Most vegetables are available in a number of varieties, which should be noted carefully before purchasing seed or plants. The days to maturity of different varieties of the same vegetable often differ, an important fact especially for gardeners in areas with short growing seasons. Some cucumbers are designed for slicing, others for pickling. Some tomatoes are best in salads, whereas others lend themselves better to processing. Flavor can vary with variety, and in many cases so can disease resistance.

You can read about specific varieties in the chapter on individual vegetables, which starts on page 69, and also on seed packets and in mail-order catalogs.

Many vegetable varieties are hybrids. Although hybrids are more expensive to buy, because of the costs of research and production of the seed, they are almost always worth the additional cost. Hybrid plants may be more vigorous, may set fruit earlier, may be more disease resistant, or may have all of these features.

*'Pixie Hybrid' tomatoes grow and flourish in small biodegradable pots.*

# Planting and Harvesting

*Here is the information that you need to get your plants off to a good start. Planting charts, tips on starting seeds indoors, data on germination periods, and advice on protecting transplants will help you every step of the way.*

Most vegetables are annuals, and most are started from seed. You don't need to plant from seed, however. You can buy many types of vegetables already started as transplants. They are sold at the right size for planting in the garden. It's easiest and quickest to buy transplants from the nursery. They cost more than seeds, however, and the variety available isn't very wide.

Most gardeners start some plants from seed and buy a few ready to go into the ground. If you want only one tomato plant for a few salad tomatoes, there's no need to buy a packet of 50 seeds. But if you're planting a large garden, the cost of transplants can be quite high. Also, if you browse through a comprehensive seed catalog, you will see that you can choose from dozens of varieties of almost any vegetable you might care to grow. It might be fun to try blue potatoes this year, or you might like to experiment with an extra-early corn to see how early in the season you can harvest fresh corn. Unusual vegetables such as these are usually only available in seed form through the mail.

*Seeds are the most economical way of obtaining plants for the vegetable garden. The seeds of some plants can be sown directly where the plant will grow. Others must be started indoors and then transplanted into the garden.*

## GROWING FROM SEED

Like most of the other gardening arts, raising plants from seed is a matter of giving the plant what it needs to grow. Seeds need a certain environment to germinate, and seedlings need slightly different conditions to grow into strong transplants. If you provide the right conditions, the rest happens automatically.

### Germination

Temperature, moisture, and oxygen supply are the three most significant factors that influence germination. In some cases, light constitutes a fourth factor. Most vegetable seeds will tolerate quite a bit of variation in these factors; however, as extremes are approached (above or below optimum), the germination rate slows, increasing the number of abnormal seedlings and reducing total germination.

**Temperature** This factor affects vegetable seed germination both directly and indirectly. Above the maximum and below the minimum germination temperatures, germination does not occur. See the chart on this page, which summarizes the relationship between soil temperature and germination for the most common vegetables.

The chart on pages 63 to 65 also outlines the soil temperature requirements for seeds. Most seeds that require cool soil to germinate should be sown directly into the garden in early spring. Seeds that must have warm soil should not be sown outdoors until the danger of frost has passed and the soil is between 50° and 60° F. When starting these seeds indoors, bottom heat may be necessary to warm the growing medium. Even though the germinating room may be the correct temperature, the medium will be cooler because of the evaporation of surface moisture. Bottom heat can be achieved by setting seed flats on a warm surface, such as the top of a refrigerator, water heater, or television set, or by using soil heating cables.

**Moisture** The stored food in a seed occurs in a very concentrated, complex form. Before it can be used, a series of chemical reactions must take place, and for this to happen, moisture must be available. Water serves two functions: It triggers the necessary series of reactions within the seed and it softens and weakens the seed coat, thus permitting the growing embryo to break through.

If there is insufficient water, germination will not occur or the seedlings will dry out and die; too much water will cause the seeds to rot. It is important to keep the humidity high around seed flats until the seeds germinate to eliminate the need to water and accidentally dislodge small seeds.

The relationship between the optimum moisture conditions and germination of some common vegetable seeds is outlined in the chart on the next page.

**Oxygen supply** Since a seed is in a state of suspended animation, its energy requirements are low. As its growth begins, triggered by the presence of moisture, oxygen combines with the stored food to produce energy.

### *Soil Temperature and Germination*

| Vegetable | Minimum | Optimum | Maximum |
|---|---|---|---|
| Asparagus | 50° F | 75° F | 95° F |
| Bean, lima | 60° F | 80° F | 85° F |
| Bean, snap | 60° F | 85° F | 95° F |
| Beet | 40° F | 85° F | 95° F |
| Broccoli | 40° F | 85° F | 95° F |
| Cabbage | 40° F | 85° F | 95° F |
| Carrot | 40° F | 80° F | 95° F |
| Cauliflower | 40° F | 80° F | 95° F |
| Celery | 40° F | 70° F | 75° F |
| Chard, Swiss | 40° F | 85° F | 95° F |
| Corn, sweet | 50° F | 85° F | 105° F |
| Cucumber | 60° F | 95° F | 105° F |
| Endive | 32° F | 75° F | 75° F |
| Lettuce | 32° F | 75° F | 75° F |
| Muskmelon | 60° F | 95° F | 105° F |
| Okra | 60° F | 95° F | 105° F |
| Onion | 32° F | 80° F | 95° F |
| Parsnip | 32° F | 70° F | 85° F |
| Pea | 40° F | 75° F | 85° F |
| Pepper | 60° F | 85° F | 95° F |
| Pumpkin | 60° F | 95° F | 105° F |
| Radish | 40° F | 85° F | 95° F |
| Spinach | 32° F | 70° F | 75° F |
| Squash | 60° F | 95° F | 105° F |
| Tomato | 50° F | 85° F | 95° F |
| Turnip | 40° F | 85° F | 105° F |
| Watermelon | 60° F | 95° F | 105° F |

Chart adapted from Harrington and Minges, University of California, Davis.

Lack of oxygen—the cause of most vegetable seeds' failure to germinate—occurs when the pore space in the soil or medium around the seed is saturated with water, such as after a heavy rain or overwatering. If flooded conditions persist for too long, the seed will rot.

Some seeds, particularly squash, pumpkin, cucumber, gourds, and related plants, are more sensitive to low oxygen than others. On the other hand, celery seed can germinate even if it is completely immersed in water.

## Moisture and Seed Germination

Some vegetable seeds germinate better with different moisture conditions than others. This chart shows how moist the soil should be for each type of seed.

| | |
|---|---|
| Germinate well in a wide range of soil-moisture conditions: | Cabbage, carrot, corn (sweet), cucumber, muskmelon, onion, pepper, radish, squash, tomato, turnip, watermelon |
| Germinate best in fairly moist soil: | Beet, endive, lettuce, lima bean, pea, snap bean |
| Germinate only in wet soil: | Celery |
| Germinate best in relatively dry soil: | Spinach, spinach (New Zealand) |

Chart adapted from Doneen and MacGillvray.

*Pumpkin, squash, cucumber, and melon seeds are often planted in hills, or clusters, rather than rows.*

**Light** Seedlings need strong light to grow into healthy plants that are not too tall or spindly. Indoors, they require a south window or fluorescent lights to have enough light to photosynthesize (convert water and carbon dioxide from the air into food). The warmer the location, the brighter the light need be. The best situation for raising most vegetable seedlings is a slightly cool temperature (below 75° F) and full sun (or bright fluorescent light) for at least six hours a day.

## STARTING SEEDS INDOORS

Although the seeds of vegetables such as root crops, beans, peas, and corn should be sown directly where you want them to grow, some vegetable seeds, such as broccoli, brussels sprouts, eggplant, celery, tomatoes, peppers, and head lettuce, grow more successfully when started indoors. Some seeds can be started either way. The reasons for starting seeds indoors are varied: The growing season is so long that plants won't produce unless given a head start; plants will be more productive if the growing season is lengthened; seeds are small and might wash away if sown outdoors. Plants begun as transplants also avoid some of the hazards common to seedlings outdoors—birds, insects, and weeds.

There are different methods of growing from seed to transplant size. Some of them are illustrated below.

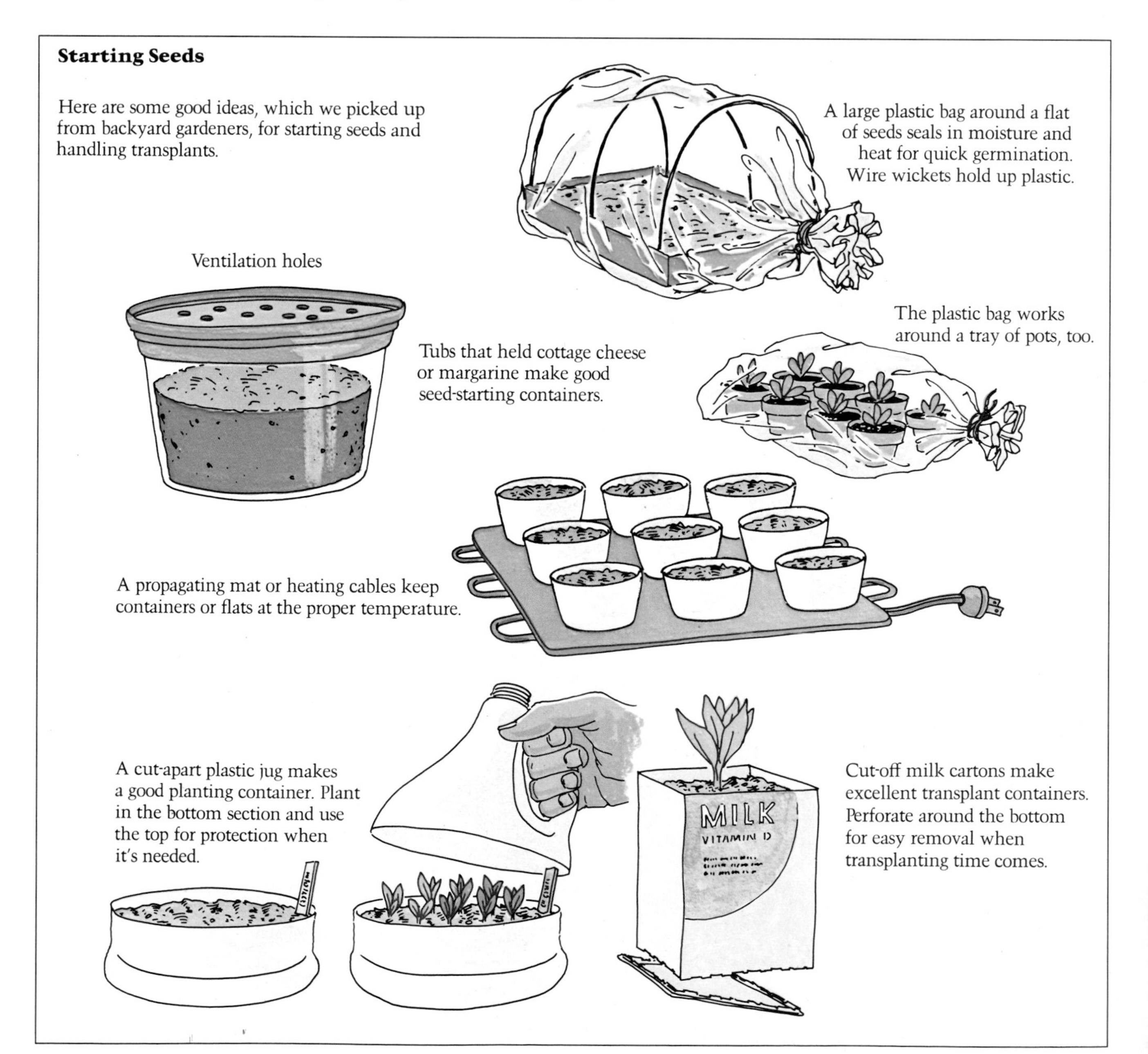

To grow successful transplants, you need:

- ☐ Suitable containers,
- ☐ A disease-free growing medium,
- ☐ Warmth and moisture for seed germination,
- ☐ Adequate light for stocky growth, and
- ☐ An adjustment period to ready the indoor plants to outdoor conditions.

### Containers

Anything that can hold growing medium, is the right size, and has proper drainage can be used to germinate seeds.

**Seed flats** The traditional method of starting seeds is to germinate them in a flat, then transplant the seedlings, when they have grown enough, into individual pots. Seed flats can be of any length and width and should be 3 to 3½ inches deep for good root development. They can be homemade or purchased; store-bought ones are usually plastic or compressed fiber. Fiber flats cannot be reused since they cannot be sterilized after use. They also dry out more quickly, although they have better aeration.

If you make your own flats or reuse plastic flats, be sure to wash them well and rinse them in a 10 percent bleach solution to prevent the spread of diseases.

**Individual containers** Although fiber and plastic pots can be used for transplanting, all of the below are ideal for sowing and growing in one step: Because transplanting is eliminated the roots do not need to be disturbed. These are perfect for large seeds. Several kinds are available.

*Fiber pots* These containers are made of peat or other fibrous material. Fill them with soilless growing mix and place seeds in the mix. Once the seedlings are mature, set the whole container out in the garden or into a planting container. Plastic pots can be used in the same way as fiber pots but of course are not plantable.

*Compressed peat pellets* These pellets are pressed from peat moss and contain fertilizer. When placed in water they expand to make a 1¾- by 2-inch container. After they have expanded, place them directly into a tray to hold them.

*Plugs* A new method of seed germination is known as plug growing. Plugs are seedlings that have cone-shaped or cylindrical rootballs. Plug trays, with up to 200 openings, are filled with soilless growing mix and one or two seeds are sown into each compartment. As with the other methods, seedlings go from germination into the garden without the need for transplanting.

**Soil blocks** Another new technique presses the potting mix itself into a block that holds its shape. A special tool is needed for making

**Seed-Starting Containers**

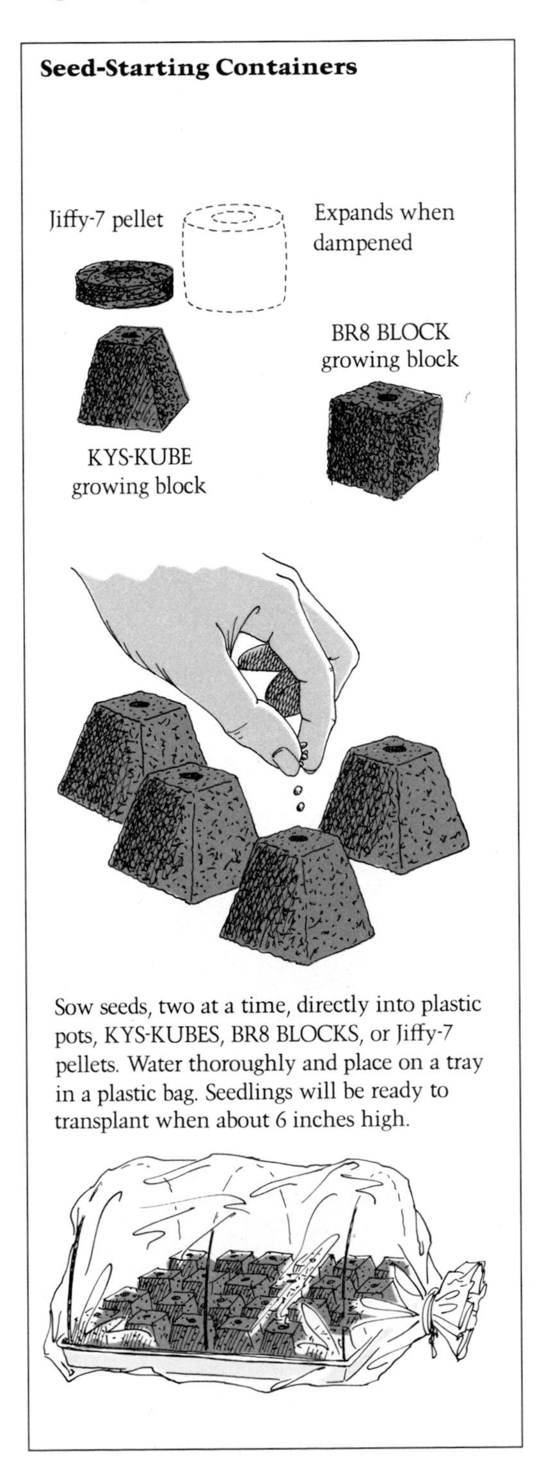


Sow seeds, two at a time, directly into plastic pots, KYS-KUBES, BR8 BLOCKS, or Jiffy-7 pellets. Water thoroughly and place on a tray in a plastic bag. Seedlings will be ready to transplant when about 6 inches high.

*Special tools for making soil blocks are available from many seed companies and garden supply outlets.*

the blocks, and special techniques are needed for watering them. (Directions for growing seedlings in soil blocks are included with the block-making device.) Since this method uses no containers at all, it is very economical once the block maker is purchased.

## Germinating Medium

The best medium for germinating seeds is a sterile, soilless mix of peat or sphagnum moss with perlite and/or vermiculite. You can make the mix yourself in a 50:50 ratio, or purchase mix that is ready to use. (For more information on soilless mixes, see page 17.) Mixes for starting seeds should be finer than those for growing plants in containers. Many purchased mixes have fertilizer added to give the seedlings a boost, and a wetting agent in order to improve water retention. It is important not to reuse the mix for seed germination, because it will not be sterile and may carry diseases, but it may be used for transplanting or as a container mix for other plants.

Seeds may be started in pure vermiculite. The seedlings will be easy to transplant and less prone to damping-off disease, which causes young seedlings to suddenly whither at the soil line, topple over, and die.

Starting seeds in garden soil is not recommended, since it usually does not have the proper drainage and aeration for optimum germination. Soil also carries disease, insects, and weed seeds.

## Sowing

The first step in sowing seed is to premoisten the medium, and the containers if they are made of fiber or peat. Fill the containers to within ¼ inch of the top with medium and water them with a solution of ½ tablespoon benomyl per gallon of water to prevent damping-off disease. Allow excess water to drain, then sow the seed.

Before you start, refer to the chart on pages 63 to 65; see Weeks Needed to Grow to Transplant Size. Count backward from the intended outdoor planting date to double-check that the time is right to start seeds indoors. Cucumbers need only 4 weeks to reach outdoor transplanting size, but celery needs 10 to 12 weeks.

Sow one or two large seeds into individual containers; sow small seed in rows in flats to make transplanting easier. Large seeds can be placed by hand, smaller ones by tapping the seed packet. Don't sow seeds too close; seedlings need room for root growth, light, and air, so allow at least the diameter of the seed between seeds. Small seeds should not be covered but pressed into the surface of the growing medium; others should be covered with medium to a thickness of one to two times their diameter.

Place individual containers or flats in a plastic bag and set them where the temperature is right for germination. Once seedlings pop through the medium, remove the plastic bag.

### Sowing Carrots

To make the most of your space when sowing carrot seed, sow randomly in a swath 6 to 12 inches wide, and cover with ¼ inch of fine peat moss. Thin seedlings randomly and enjoy the sweet, tender miniature carrots as they grow.

### Sowing Small Seeds

When the small size or color of seeds makes them difficult to see as you're sowing, lay sheets of tissue paper in the trench. The tissue will decompose quickly when covered and watered. Seed tape is also available for many plants.

### Spacing Seeds

Space small seeds evenly by rubbing a pinch between fingers or

tap them directly from the packet.

### Trench Greenhouse

To help start hard-to-germinate seeds such as tomatoes, peppers, and eggplant, plant seeds in a trench covered with clear plastic. Angle it so water drains off.

*Before transplanting a seedling, water it well to be sure the roots get wet. In the moistened, amended soil, dig a hole slightly larger than the rootball of the plant. Place the plant in the hole and press the soil around it.*

**Transplanting**

1. Ready the plant. If it's in a peat pot, tear off the top edge so it can't act as a wick and dry out the rootball.

2. If it's in a plastic, fiber, or clay pot, tip it out; don't pull it out by the stem.

3. After planting, firm the soil around the transplant. Then water lightly to settle the soil and remove any air pockets that may be left around the rootball.

4. To make sure the rootball stays moist during the first few critical days, build a small temporary basin, a little larger than the root system.

## Caring for Seedlings

Once seedlings have germinated, keep the medium evenly moist but not wet. Watering from the bottom is best when seedlings are small, so they will not be dislodged. This is also the time to move them into full sunlight—12 hours a day, if possible. If you can't find enough sun, grow the seedlings 3 to 6 inches high under fluorescent lights. Daytime temperatures should range from 70° to 75° F; nighttime temperatures should be between 60° and 65° F.

When the first true leaves (the first "seed leaves" don't look like mature leaves) have developed, start fertilizing with quarter-strength soluble fertilizer every week, increasing to half strength as the plants grow.

When seedlings grown in flats have formed two sets of true leaves above the seed leaves, dig them out carefully and transplant them into containers filled with soilless mix. With a pencil make a small hole in the mix, then set the seedling into the hole; the seed leaves should be ½ inch above the surface. Press the mix firmly around the roots and stem. Water carefully.

## Hardening Off Transplants

Don't set young plants directly into the open garden from an indoor environment. Instead, starting about one week before transplanting them into the garden, take them outside in the daytime and bring them in again at night if frost is likely. Gradually expose them to lower temperatures and more sunlight.

*A sunshade protects tender lettuce plants during warm weather.*

## Setting Out

Transplants should go into the garden with a minimum of root disturbance. Water the ground and the transplants prior to planting to reduce shock, and transplant late in the afternoon or on a cloudy day. If you have used pots that aren't plantable, turn the pots upside down and tap them on the bottom to remove the rootballs without disturbing them.

Dig a hole twice the size of the rootball, set the transplant in place, and firm the soil around the rootball. With peat pots, cubes, and blocks, there will be very little disturbance of the roots; however, to prevent the rootball from drying out rapidly, place all such containers below soil level and peel away the outer layer of peat.

In hot weather give the transplants supplemental water between the regular irrigations.

Young transplants will mostly need protection from stress. If a late frost threatens, cover the transplants with hot caps or plastic film. When using hot caps or plastic covers, make sure to provide some ventilation so that the young plants won't be cooked by the heat buildup. Coffee cans and plastic foam cups with the tops and bottoms removed can also be used for frost protection, without heat buildup. Also refer to Extending the Season on page 28 for additional suggestions.

Young plants may need protection from wind. This can be done with a screen of hardware cloth or burlap.

If the sun is intense, shade the plants for a few days with pieces of wood or shingles to give the plants a chance to get used to their brighter environment. See pages 34 and 35 for ways to keep cutworms, snails, and slugs from feasting on your vegetables.

To confirm the best time to set out transplants, refer to the chart on pages 63 to 65 and to the descriptions of individual vegetables starting on page 69.

## STARTING SEEDS OUTDOORS

The first step in sowing seeds outdoors is preparing the soil; it must be friable, and fertile for most plants to grow properly. Follow the tips on page 12 for soil preparation. Next, check the garden calendar on page 43, the descriptions of individual vegetables starting on page 69, and the chart on pages 63 to 65 to be sure that you are sowing seeds at the right time.

Before you sow, water the soil thoroughly. Then, with a yardstick or the corner of a board, make shallow furrows in the prepared soil, about the thickness of the seed to be planted.

*Scrap wood is used to shield a young tomato plant from the hot sun.*

Sow small seeds thinly and evenly by tapping the seed packet with a pencil or your finger. Large seeds can be placed by hand. To allow for seeds that will not germinate or will be eaten by birds, sow seeds about twice as close as their final spacing should be. After sowing, cover small seeds with sand or potting mix; pull the soil over larger seeds. Firm the soil with a board or the back of a rake. Seeds need close contact with the soil in order to germinate properly.

Proper sowing depth is critical to good germination and growth. If the seeds are planted too shallow, they may dry out before they germinate. If they are planted too deep, they may not germinate or grow because the soil could be too cold at the lower depth, there could be insufficient oxygen for the seeds, the seeds may have exhausted their food supply and died before they reached the surface, or the seeds may not have been strong enough to push through the soil.

Take into consideration weather and soil conditions when calculating sowing depth. If the weather is wet or the soil heavy, plant shallow. If the soil is sandy and dry weather is expected, plant a little deeper than normal.

After sowing is completed, water the seedbed with a very fine spray. Check every day to see if it needs to be watered, and never let the soil dry out. After the seeds germinate, gradually reduce the watering frequency as the plants mature. This will actually encourage better and stronger root growth, because the roots will grow deeper into the soil looking for moisture.

If sowing is done in the heat of summer, some sort of shading will be needed to slow evaporation and protect the seedlings from strong sunlight. Burlap and cheesecloth are both effective.

Keep the area well weeded, since weeds will compete with the seedlings for food and water and will cut down on air circulation, which can lead to disease. Remove the weeds carefully so you don't disturb the seedlings' roots, and water after weeding just in case the roots have been jostled.

Be on the lookout for insects and diseases and refer to pages 32 to 37 if you encounter any problems with them.

After the seedlings have reached a height of 2 to 3 inches, it's time to thin them. Follow the guidelines on the seed packet or the chart on pages 63 to 65 to determine the proper distance that should be left between seedlings.

*A piece of corrugated fiberglass roofing forms a miniature greenhouse that allows these seedlings to be set out weeks earlier than would be otherwise possible.*

*This raised bed with a fiberglass cover allows sweet potato seedlings to thrive in a cooler-than-usual climate.*

Thinning is best done after a rain or watering. If you can't bear to throw away the thinnings, move them to other parts of the garden or give them away. Some thinnings, such as lettuce and beets, are good to eat.

### Sowing Small Seeds

Space small seeds evenly by rubbing a pinch of them between your fingers or tap them directly from the seed packet. When the small size or color of seeds makes them difficult to see as you're sowing, lay sheets of tissue paper in the trench. The tissue will decompose quickly when covered and watered. Seed tape is also available for many plants.

### Trench Greenhouse

To help start hard-to-germinate seeds, such as tomatoes, peppers, and eggplant, plant seeds in a trench covered with clear plastic. Angle the plastic so water drains off. The plastic forms a tiny greenhouse that provides the warmth these seeds need to germinate well.

### Seed Storage

Some seed, such as parsnip, is short-lived and cannot be saved from one year to the next. Most seed that is stored under ideal conditions will maintain high germination rates, even up to several times longer than most seed-storage information suggests. However, if seed is stored improperly, it may die within a month.

Ideal seed-storage conditions are basically the reverse of those for ideal seed germination. The two most important factors are the moisture content of the seed and the temperature. If the moisture content of the seed is low, the seed will keep well under a wide range of temperatures. If the seed is both dry and cold, then conditions are nearly ideal.

Seed packets that are made of or lined with aluminum foil and have not been opened can be stored from one year to the next if they are kept between 32° and 40° F. Seeds in paper packets and packets that have been opened should be placed in a plastic bag with silica gel to absorb moisture, and kept cool.

To test stored seed to see if it is still good, wrap 10 seeds in a moist paper towel and place it in a plastic bag. After the normal number of days has passed for the seeds to germinate, open the towel. If fewer than half have sprouted, you should not use the seeds. If 5 to 8 have sprouted, use them but sow more heavily than usual. If 8 to 10 seeds have sprouted, sow them at the usual rate.

*If transplants of certain varieties aren't available, you may have to start from seed. Pepper seeds take as long as eight weeks to germinate and grow to transplant size. Another month or two of growth outdoors is required before fruit is ready for harvest.*

## PURCHASING TRANSPLANTS

The gardener who does not have the facilities or the desire to start vegetables from seed can purchase transplants, commonly called bedding plants, at local garden centers and nurseries. They have been raised to be available at the proper planting time for your area. Look for healthy plants with deep green foliage that have no evidence of insects or diseases. Avoid tall, spindly plants, stunted plants, and those whose care leaves a question in your mind.

As with buying seeds it's important to select varieties that will do best in your garden. Don't purchase plants too much in advance of when you intend to plant them, since care will need to be given to them in the interim and most garden centers and nurseries restock once or twice a week. Avoid plants that are already in bloom, since plants that are not flowering will actually reach maturity faster and be more satisfactory than ones planted in bloom.

If you can't plant the new vegetables right after you've bought them, place them outside in a partially shaded spot and check every day to see if they need watering. To plant, follow the directions in Setting Out on pages 58 and 59.

## HOW TO USE THE PLANTING CHART

The planting chart on pages 63 to 65 provides all the information you will need to add vegetables, whether grown from seed or from transplants, to the garden.

**Depth to plant seed** A quick look at the figures in this column will tell you that many gardeners plant too deep.

**Number of seeds to sow per foot** This figure tells you how far apart to sow seeds.

**Distance between plants** These figures, given in inches, are the optimum final spacing between plants. When two figures are given, the first figure is the minimum spacing; wider spacing produces better growth.

**Distance between rows** The minimum distance assumes that space is limited and weeding will be done with hand tools. Wider distance between rows is preferable.

**Number of days to germination** These figures indicate how soon you can expect to see signs of life. The number of days varies by soil temperature. Early spring sowings will take longer than later plantings.

**Soil temperature for seed** Seeds that require cool soil do best in a temperature range of 50° to 65° F, which occurs in most areas in early spring as soon as the soil can be worked; seeds that tolerate cool soil do best in a 50° to 85° F range, which occurs from midspring on, starting about four weeks before the last frost; those that require warm soil do best in a 65° to 85° F range, which occurs after the last spring frost. These same guidelines are to be used to time the setting out of transplants.

**Weeks needed to grow to transplant size** These figures indicate the number of weeks the seeds need indoors or in a greenhouse after sowing until the plants are large enough to be moved to the garden.

**Days to maturity** Figures in this column show the relative length of time needed to grow a crop from seed or transplant to table use. The time will vary by variety and season.

## *Planting Chart*

| Vegetable | Depth to Plant Seed | Number of Seeds to Sow per Foot | Distance Between Plants | Distance Between Rows | Number of Days to Germination | Needs Cool Soil | Tolerates Cool Soil | Needs Warm Soil | Weeks Needed to Grow to Transplant Size | Days to Maturity | Remarks |
|---|---|---|---|---|---|---|---|---|---|---|---|
| Artichoke | ½″ | | 60″ | 72″ | 7–14 | | • | | 4–6* | 12 mos. | Starting with divisions preferred. |
| Arugula | ¼″ | 8-10 | 8″–12″ | 18″–24″ | 7–14 | | | • | | 60 | |
| Asparagus | 1½″ | | 18″ | 36″ | 7–21 | | • | | 12–14* | 3 years | Sow in spring and transplant the following spring. |
| Asparagus bean, or yard-long bean | ½″–1″ | 2-4 | 12″–24″ | 24″–36″ | 6–13 | | | • | | 65–80 | Variety of black-eyed peas. Grow as pole beans. |
| Bean | | | | | | | | | | | |
| Snap bush | 1″–1½″ | 6-8 | 2″–3″ | 18″–30″ | 6–14 | | • | | | 45–65 | Make sequence plantings. |
| Snap pole | 1″–1½″ | 4-6 | 4″–6″ | 36″–48″ | 6–14 | | | • | | 60–70 | Long bearing season if kept picked. |
| Lima bush | 1½″–2″ | 5-8 | 3″–6″ | 24″–30″ | 7–12 | | | • | | 60–80 | Needs warmer soil than snap beans. |
| Lima pole | 1½″–2″ | 4-5 | 6″–10″ | 30″–36″ | 7–12 | | | • | | 85–90 | |
| Fava | | | | | | | | | | | |
| Broad bean | | | | | | | | | | | |
| Windsor bean | 2½″ | 5-8 | 3″–4″ | 18″–24″ | 7–14 | | • | | | 80–90 | Hardier than common beans. |
| Garbanzo, or Chick-pea | 1½″–2″ | 5-8 | 3″–4″ | 24″–30″ | 6–12 | | | • | | 105 | |
| Scarlet runner | 1″–1½″ | 4-6 | 4″–6″ | 36″–48″ | 6–14 | | | • | | 60–70 | Will grow in cooler summers than common beans. |
| Soybean | 1½″–2″ | 6-8 | 2″–3″ | 24″–30″ | 6–14 | | | • | | 55–85, 95–100 | |
| Beet | ½″–1″ | 10-15 | 2″ | 12″–18″ | 7–10 | | • | | | 55–65 | Thin out extra plants and use for greens. |
| Black-eyed pea, cowpea, or southern pea | ½″–1″ | 5-8 | 3″–4″ | 24″–30″ | 7–10 | | | • | | 65–80 | |
| Broccoli | ½″ | 10-15 | 14″–18″ | 24″–30″ | 3–10 | | • | | 5–7* | 60–80T | 80–100 days from seed. |
| Brussels sprouts | ½″ | 10-15 | 12″–18″ | 24″–30″ | 3–10 | | • | | 4–6* | 80–90T | 100–110 days from seed. |
| Cabbage | ½″ | 8-10 | 12″–20″ | 24″–30″ | 4–10 | | • | | 5–7* | 65–95T | Use thinnings for transplants. 90–150 days from seed. |
| Cabbage, Chinese | ½″ | 8-10 | 10″–12″ | 18″–24″ | 4–10 | | • | | 4–6 | 80–90 | Best as seeded fall crop. |
| Cardoon | ½″ | 4-6 | 18″ | 36″ | 8–14 | | • | | 8 | 120–150 | Transplanting to harvest about 90 days. |
| Carrot | ¼″ | 15-20 | 1″–2″ | 14″–24″ | 10–17 | | • | | | 60–80 | Start using when ½ in. diameter to thin stand. |
| Cauliflower | ½″ | 8-10 | 18″ | 30″–36″ | 4–10 | | • | | 5–7* | 55–65T | 70–120 days from seed. |
| Celeriac | ⅛″ | 8-12 | 8″ | 24″–30″ | 9–21 | • | | | 10–12* | 90–120T | Keep seeds moist. |
| Celery | ⅛″ | 8-12 | 8″ | 24″–30″ | 9–21 | • | | | 10–12* | 90–120T | Keep seeds moist. |
| Celtuce | ½″ | 8-10 | 12″ | 18″ | 4–10 | | • | | 4–6 | 80 | Same culture as lettuce. |
| Chard, Swiss | 1″ | 6-10 | 4″–8″ | 18″–24″ | 7–10 | | • | | | 55–65 | Use thinnings for early greens. |
| Chayote | See text | | 10 ft. | | See text | | | • | | Peren | Plant whole fruit. |
| Chicory, witloof, or Belgian endive | ¼″ | 8-10 | 4″–8″ | 18″–24″ | 5–12 | | • | | | 90–120 | Force mature root for Belgian endive. |
| Chives | ½″ | 8-10 | 8″ | 10″–16″ | 8–12 | | • | | 6–8 | 80–90 | Also propagate by division of clumps. |
| Collards | ¼″ | 10-12 | 10″–15″ | 24″–30″ | 4–10 | | • | | 4–6* | 65–85T | Direct seed for a fall crop. |
| Corn, sweet | 2″ | 4-6 | 10″–14″ | 30″–36″ | 6–10 | | | • | | 60–90 | Make successive plantings. |
| Cress, garden | ¼″ | 10-12 | 2″–3″ | 12″–16″ | 4–10 | | • | | | 24–45 | Seeds sensitive to light. |
| Cucumber | 1″ | 3-5 | 12″ | 48″–72″ | 6–10 | | | • | 4 | 55–65 | See text about training. |

* Transplants preferred over seed. T = Number of days from setting out transplants; all others are from seeding.

# *Planting Chart (continued)*

| Vegetable | Depth to Plant Seed | Number of Seeds to Sow per Foot | Distance Between Plants | Distance Between Rows | Number of Days to Germination | Needs Cool Soil | Tolerates Cool Soil | Needs Warm Soil | Weeks Needed to Grow to Transplant Size | Days to Maturity | Remarks |
|---|---|---|---|---|---|---|---|---|---|---|---|
| Dandelion | ½″ | 6-10 | 8″-10″ | 12″-16″ | 7-14 | | • | | | 70-90 | |
| Eggplant | ¼″-½″ | 8-12 | 18″ | 36″ | 7-14 | | | • | 6-9* | 75-95T | |
| Endive | ½″ | 4-6 | 9″-12″ | 12″-24″ | 5-9 | | • | | 4-6 | 60-90 | Same culture as lettuce. |
| Fennel, Florence | ½″ | 8-12 | 6″ | 18″-24″ | 6-17 | | • | | | 120 | Plant in fall in mild-winter areas. |
| Garlic | Bulb, 1″ | | 2″-4″ | 12″-18″ | 6-10 | | • | | | 90, sets | |
| Gourd | | | | | | | | | | | See text. |
| Ground-cherry, or husk tomato | ½″ | 6 | 24″ | 36″ | 6-13 | | | • | 6* | 90-100T | Treat same as tomatoes. |
| Horseradish | Div. | | 10″-18″ | 24″ | | | • | | | 6-8 mos. | Use root division 2-8 in. long. |
| Huckleberry, garden | ½″ | 8-12 | 24″-36″ | 24″-36″ | 5-15 | | | • | 5-10 | 60-80 | |
| Jerusalem artichoke | Tubers, 4″ | | 15″-24″ | 30″-60″ | | | • | | | 100-105 | |
| Jicama | ¼″ | | 6″-8″ | | 7 | | | • | 2 | 4-8 mos. | Seeds costly, start indoors in peat pots. |
| Kale | ½″ | 8-12 | 8″-12″ | 18″-24″ | 3-10 | | • | | 4-6 | 55-80 | Direct seed for fall crop. |
| Kohlrabi | ½″ | 8-12 | 8″-12″ | 18″-24″ | 3-10 | | • | | 4-6 | 60-70 | |
| Leek | ½″-1″ | 8-12 | 2″-4″ | 12″-18″ | 7-12 | | • | | 10-12* | 80-90T | 130-150 days from seed. |
| Lettuce, head | ¼″-½″ | 4-8 | 12″-14″ | 18″-24″ | 4-10 | • | | | 3-5* | 55-80 | Keep seeds moist. |
| Lettuce, leaf | ¼″-½″ | 8-12 | 4″-6″ | 12″-18″ | 4-10 | • | | | 3-5 | 45-60 | Keep seeds moist. |
| Muskmelon | 1″ | 3-6 | 12″ | 48″-72″ | 4-8 | | | • | 3-4 | 75-100 | |
| Mustard | ½″ | 8-10 | 2″-6″ | 12″-18″ | 3-10 | | • | | | 40-60 | Use early to thin. |
| Nasturtium | ½″-1″ | 4-8 | 4″-10″ | 18″-36″ | | | • | | | 50-60 | |
| Okra | 1″ | 6-8 | 15″-18″ | 28″-36″ | 7-14 | | | • | 4-6 | 50-60 | |
| Onion | | | | | | | | | | | |
| Sets | 1″-2″ | | 2″-3″ | 12″-24″ | | • | | | | 95-120 | Green onions 50-60 days. |
| Plants | 2″-3″ | | 2″-3″ | 12″-24″ | | • | | | 8 | 95-120T | |
| Seed | ½″ | 10-15 | 2″-3″ | 12″-24″ | 7-12 | • | | | | 100-165 | |
| Parsnip | ½″ | 8-12 | 3″-4″ | 16″-24″ | 15-25 | | • | | | 100-120 | |
| Pea | 2″ | 6-7 | 2″-3″ | 18″-30″ | 6-15 | • | | | | 65-85 | |
| Peanut | 1½″ | 2-3 | 6″-10″ | 30″ | | | | • | | 110-120 | Requires warm growing season. |
| Pepper | ¼″ | 6-8 | 18″-24″ | 24″-36″ | 10-20 | | | • | 6-8* | 60-80T | |
| Potato | 4″ | 1 | 12″ | 24″-36″ | 8-16 | | • | | | 90-105 | |
| Pumpkin | 1″-1½″ | 2 | 30″ | 72″-120″ | 6-10 | | | • | | 70-110 | Give them room. |
| Radicchio | ¼″ | 4-5 | 8″-10″ | 12″ | 5-12 | | • | | | 80-110 | |
| Radish | ½″ | 14-16 | 1″-2″ | 6″-12″ | 3-10 | • | | | | 20-50 | Early spring or late fall weather. |
| Rhubarb | Crown | | 24″-30″ | 36″ | | | • | | 1 yr. | 2 yrs. | Matures second season. |
| Rutabaga | ½″ | 4-6 | 8″-12″ | 18″-24″ | 3-10 | | • | | | 80-90 | |
| Salsify | ½″ | 8-12 | 2″-3″ | 16″-18″ | | • | | | | 110-150 | |
| Salsify, black | ½″ | 8-12 | 2″-3″ | 16″-18″ | | • | | | | 110-150 | |
| Shallot | Bulb, 1″ | | 2″-4″ | 12″-18″ | | | • | | | 60-75 | |
| Shungiku | ½″ | 15-20 | 2″-3″ | 10″-12″ | 5-14 | | • | | | 42 | Best in cool weather. |
| Sorrel | ⅛″ | 4 | 10″ | 12″ | 10 | • | | | | 100 | |

* Transplants preferred over seed. T = Number of days from setting out transplants; all others are from seeding.

## Planting Chart (continued)

| Vegetable | Depth to Plant Seed | Number of Seeds to Sow per Foot | Distance Between Plants | Distance Between Rows | Number of Days to Germination | Needs Cool Soil | Tolerates Cool Soil | Needs Warm Soil | Weeks Needed to Grow to Transplant Size | Days to Maturity | Remarks |
|---|---|---|---|---|---|---|---|---|---|---|---|
| Spinach | ½″ | 10-12 | 2″-4″ | 12″-14″ | 6-14 | • | | | | 40-65 | |
| Malabar | ½″ | 4-6 | 12″ | 12″ | 10 | • | | | 6-8* | 70 | |
| New Zealand | 1½″ | 4-6 | 18″ | 24″ | 5-10 | | • | | | 70-80 | |
| Tampala | ¼″-½″ | 6-10 | 4″-6″ | 24″-30″ | | | • | | | 21-42 | Thin; use early while tender. |
| Squash | | | | | | | | | | | |
| Summer | 1″ | 4-6 | 16″-24″ | 36″-60″ | 3-12 | | | • | 3-4 | 50-60 | |
| Winter | 1″ | 1-2 | 24″-48″ | 72″-120″ | 6-10 | | | • | 3-4 | 85-120 | |
| Sunflower | 1″ | 2-3 | 16″-24″ | 36″-48″ | 7-12 | | | • | | 80-90 | Space wide for large heads. |
| Sweet potato | Plants | | 12″-18″ | 36″-48″ | | | | • | | 120 | Propagate from cuttings. |
| Tomatillo | ½″ | 6 | 24″ | 36″ | 6-13 | | | • | 6 | 90-100T | |
| Tomato | ½″ | | 18″-36″ | 36″-60″ | 6-14 | | | • | 5-7* | 55-90T | Early var. 55-60. Mid 65-75. Late 80-100. |
| Turnip | ½″ | 14-16 | 1″-3″ | 15″-18″ | 3-10 | • | | | | 45-60 | Thin early for greens. |
| Watermelon | 1″ | | 12″-16″ | 60″ | 3-12 | | | • | | 80-100 | Icebox size matures earlier. |

* Transplants preferred over seed. T = Number of days from setting out transplants; all others are from seeding.

## HARVESTING

Most gardeners enjoy planting and harvesting. Whereas much of the rest of gardening is controlling the environment so the plants are free to grow to their fullest potential, in planting and harvesting, gardeners handle the plants themselves.

The harvest season for some vegetables is very long; lettuce is delicious from the time it's big enough to eat until it bolts weeks later, and carrots will wait in the ground for you to pick them for weeks or even months. But other vegetables, such as peas and corn, must be picked within a day or two of their prime in order to be at their best. In the next chapter, the discussion of each vegetable tells how to know when it's ready to be harvested.

Although many vegetables are eaten fresh on the day they're picked, one of the easiest ways to extend the season during which you eat your own vegetables is to preserve them by freezing, drying, storing, canning, or some other method.

### Storing Vegetables

Under the proper conditions, many vegetables can be stored for quite a long time. A cool basement can be an excellent place to store vegetables; so can a refrigerator. Some people also build special underground storage pits.

*The edible tubers of Jerusalem artichoke, an attractive flowering perennial, should be harvested yearly.*

Some vegetables can actually be stored in the ground where they are growing. These include carrots, leeks, onions, kale, Jerusalem artichokes, and parsnips. The vegetables don't grow during the winter months, but with enough mulch to keep the ground from freezing, the crop will be kept fresh and can be harvested as needed.

Listed on the next page are the ideal storage conditions for many of the vegetables described in this book.

## Storage Recommendations

| Vegetable | Temperature (°F) | Humidity (%) | Approximate Length of Storage Period |
|---|---|---|---|
| **Cold, Moist Storage** | | | |
| Asparagus | 32-35 | 85-90 | 2-3 weeks |
| Beet, topped | 32 | 95 | 3-5 months |
| Broccoli | 32-35 | 90-95 | 10-14 days |
| Brussels sprout | 32-35 | 90-95 | 3-5 weeks |
| Cabbage, Chinese | 32 | 90-95 | 1-2 months |
| Cabbage, late | 32 | 90-95 | 3-4 months |
| Carrot, mature and topped | 32-35 | 90-95 | 4-5 months |
| Cauliflower | 32-35 | 85-90 | 2-4 weeks |
| Celeriac | 32 | 90-95 | 3-4 months |
| Celery | 32-35 | 90-95 | 2-3 months |
| Collards | 32-35 | 90-95 | 10-14 days |
| Corn, sweet | 32-35 | 85-90 | 4-8 days |
| Endive, escarole | 32 | 90-95 | 2-3 weeks |
| Greens, leafy | 32 | 90-95 | 10-14 days |
| Horseradish | 30-33 | 90-95 | 10-12 months |
| Kale | 32 | 90-95 | 10-14 days |
| Kohlrabi | 32 | 90-95 | 2-4 weeks |
| Leek, green | 32 | 90-95 | 1-3 months |
| Lettuce | 32-35 | 90-95 | 2-3 weeks |
| Onion, green | 32-35 | 90-95 | 3-4 weeks |
| Parsnip | 32-35 | 90-95 | 2-6 months |
| Pea | 35-40 | 85-90 | 1-3 weeks |
| Potato, late crop | 35-40 | 85-90 | 4-9 months |
| Radish | 32-35 | 90-95 | 3-4 weeks |
| Rhubarb | 32-35 | 90-95 | 2-4 weeks |
| Rutabaga | 32-35 | 90-95 | 2-4 months |
| Spinach | 32-35 | 90-95 | 10-14 days |
| Turnip | 32 | 90-95 | 4-5 months |

| Vegetable | Temperature (°F) | Humidity (%) | Approximate Length of Storage Period |
|---|---|---|---|
| **Cool, Moist Storage** | | | |
| Bean, snap | 40-45 | 90-95 | 7-10 days |
| Bean, lima | 32-40 | 90 | 1-2 weeks |
| Cantaloupe | 40 | 90 | 15 days |
| Cucumber | 40-50 | 85-90 | 10-14 days |
| Eggplant | 40-50 | 85-90 | 1 week |
| Okra | 45 | 90-95 | 7-10 days |
| Pepper, bell | 40-50 | 85-90 | 2-3 weeks |
| Potato, early | 50 | 90 | 1-3 weeks |
| Potato, late | 40 | 90 | 4-9 months |
| Squash, summer | 40-50 | 90 | 5-14 days |
| Tomato, ripe | 40-50 | 85-90 | 4-7 days |
| Tomato, unripe | 60-70 | 85-90 | 1-3 weeks |
| Watermelon | 40-50 | 80-85 | 2-3 weeks |
| **Cool, Dry Storage** | | | |
| Bean, dried | 32-40 | 40 | Over 1 year |
| Chile, dried | 32-50 | 60-70 | 6 months |
| Garlic, dried | 32 | 65-70 | 6-7 months |
| Onion, dried | 32 | 65-70 | 1-8 months |
| Pea, dried | 32-40 | 40 | Over 1 year |
| Shallot, dried | 32 | 60-70 | 6-7 months |
| **Warm, Dry Storage** | | | |
| Pumpkin | 55-65 | 40-70 | 2-4 months |
| Squash, winter | 55-65 | 40-70 | 3-6 months |
| Sweet potato | 55-60 | 70-85 | 4-6 months |
| Tomato, unripe | 55-70 | 85-90 | 1-3 weeks |

Chart adapted from Wright, R. C., D. H. Rose, and T. M. Whiteman, 1954. "The commercial storage of fruits, vegetables and florist and nursery stock." USDA Handbook No. 66.

### Drying Vegetables

A surprisingly large number of vegetables can be dried easily. Dried vegetables are good in soups and stews, and are a handy addition to camping trips.

Home driers are available in many garden stores, mail-order seed catalogs, and department stores. For best results, pick the vegetables in their prime, blanch them if necessary, and dry them immediately. See Ortho's book *Freezing & Drying* for more information on drying vegetables.

### Freezing Vegetables

Today, freezing is generally thought to be the best method for preserving flavor and color in vegetables. When you freeze them, choose only top quality to start with; don't bother to freeze those that are old or of questionable quality. Freeze vegetables as soon after picking as possible. Always try to pick them during the cool part of the day.

Also keep in mind that each vegetable has a specific method of preparation and blanching before it can be frozen. For more information, see Ortho's book *Freezing & Drying*.

## *Vegetables for Freezing*

| Vegetable | Suitability for Freezing | Comments |
|---|---|---|
| Asparagus | Excellent | Select young stalks with compact tips. |
| Bean, green | Good | Tendercrop and related varieties and Blue Lake varieties, either bush or pole, are preferred because of their good flavor. Also, Blue Lake varieties have a desirable thick flesh. |
| Bean, lima | Excellent | Fordhook types preferred. |
| Bean, wax | Good | |
| Beet | Fair | Better canned; select only small roots for freezing. |
| Broccoli | Excellent | |
| Cabbage | Not recommended | Preserve as sauerkraut. |
| Carrot | Fair | Select tender roots only. Can be diced and frozen with peas. |
| Cauliflower | Excellent | Also suitable for pickling. |
| Celery | Not recommended | Except for soup. |
| Chinese cabbage | Not recommended | Preserve as sauerkraut. |
| Corn, sweet | Good to Excellent | 'Jubilee', 'Seneca Chief', 'Golden Cross', and 'Silver Queen' preferred; corn on the cob frozen without blanching should be eaten in 6–8 weeks. |
| Cucumber | Not recommended | Preserve by pickling (see notes below). |
| Eggplant | Fair | Significant quality loss; suitable for casseroles. |
| Endive | Not recommended | |
| Kale | Good | Select young leaves only. |
| Kohlrabi | Fair | Significant quality loss, picks up strong flavor. |
| Lettuce | Not recommended | |
| Muskmelon | Fair | Firm-fleshed varieties are preferred; freeze small pieces; use within 3 months. |
| Mustard | Good | Select tender leaves and remove stems. |
| Onion | Fair | Freeze chopped, mature onions; significant quality loss; use in 3 months. |
| Parsnip | Fair | Significant quality loss. |
| Pea | Excellent | All large, wrinkled-seeded varieties are suitable; so are edible-pod varieties. |
| Pepper | Fair | Significant quality loss; better if frozen chopped; use in 3 months. |
| Popcorn | Not recommended | Store dry. |
| Potato | Not recommended | Store fresh at 40°–50° F. |
| Pumpkin | Not recommended | |
| Radish | Not recommended | |
| Rhubarb | Excellent | Varieties with red stalks, such as 'Canada Red', 'Valentine', and 'Ruby', preferred. |
| Spinach | Excellent | Savoy varieties are often preferred. |
| Squash, summer | Fair | Significant quality loss. |
| Squash, winter | Good | Be sure that squash is fully mature (hard rind); freeze cooked pieces or mash. |
| Swiss chard | Good | Select only tender leaves; remove midribs and stems. |
| Tomato | Fair | Better canned; freeze only juice or cooked tomatoes. |
| Turnip & rutabaga | Fair | Significant quality loss. |
| Watermelon | Fair | Freeze only as pieces; use within 3 months. |

Significant quality loss means that the product after being frozen is quite inferior to the fresh product.
For cucumber pickles, use pickling varieties if many pickles are to be made, although young slicing cucumbers are suitable for quick-method dills.
Chart adapted from Roth Klippstein and P. A. Minges, Home Garden Dept. of Vegetable Crops, Cornell University.

# The Vegetables

*The gardener's skill lies in knowing the needs of each vegetable. This chapter tells you how to fill those needs, how to harvest, and how to prepare the harvest for the table.*

A knowledge, which is the result of experience, of what conditions each vegetable needs to thrive is at the heart of vegetable gardening. Some plants do best in cool weather, others need some type of support, still others must not dry up during the period their fruit is ripening. This section gives you all that information. If you follow the directions here, you will have a successful vegetable garden. All that remains is for you to fine-tune your knowledge by observing what works in your own garden.

This chapter contains specific information about all the common vegetables and several uncommon ones, all listed in alphabetical order. Vegetables that need similar conditions or are closely related are grouped together. For example, carrots and beets are both listed under Root Crops, and collards and broccoli are grouped under Cabbage Family. You'll find a cross reference under each vegetable name telling you where the main reference is.

Many varieties of each vegetable are available, but there is space in this book to recommend only a few. Our recommendations are for those varieties that are most sure to satisfy you wherever you live, or for varieties that have some special attributes that make them notable, such as being extra early or very dwarf. Hundreds—perhaps thousands—of other varieties are available from seed racks and catalogs. By all means, try whatever plants appeal to you. A sensible way to introduce new vegetables in your garden is to plant most of the crop in a variety that has done well, but test a new variety each year.

*Bounty from the vegetable garden includes tomatoes, onions, garlic, radishes, zucchini, and peppers.*

Artichoke buds

'Green Globe' artichoke

## ARTICHOKES
### *Cynara scolymus*

Centuries before Christ the Romans were paying top prices for this thistle relative and preserving it for year-round use. It disappeared with the Roman Empire and did not come to light again until a thousand years later in southern Italy. Catherine de Médicis found artichokes in Florentine gardens and introduced them to France as gourmet delicacies. French and Spanish explorers brought them to this country.

Artichokes are large, juicy flower buds with heavy scales. Commercial artichoke production is restricted primarily to the cool, humid, moderate climate of the California coast from San Francisco south to Santa Barbara. However, a variety known as 'Creole' grows in southern Louisiana, and artichokes have also been produced by determined gardeners in northern states. It is important in cold-winter areas to mulch the crown for protection but not so heavily as to smother it.

Artichokes are perennial vegetables where winter temperatures do not drop below 20° F. Artichokes do not usually produce true from seed—you may or may not get good artichokes. Most are therefore planted from root divisions, which may be found at a local nursery or from mail-order catalogs. A full-sized clump will produce 3 or 4 plants.

If you wish to grow artichokes from seed, start seeds outdoors in spring about the time of the last frost, sowing about ½ inch deep in the garden; or start them indoors 4 to 6 weeks before the average last frost, especially where the growing season is short.

Plant root divisions 6 to 8 inches deep and allow 4 to 6 feet between plants. Leave about 7 feet between rows—the plants can get big. Plant to a side of the garden where the bed won't be in the way of the more frequent planting and cultivating of annual vegetables, and allow for some afternoon shade in hot-summer areas.

In areas of year-round production, feed plants in the fall with a high-nitrogen fertilizer and divide every 3 to 4 years to keep production high. In cold-winter areas feed in spring when new growth starts with about a pound of 10-10-10 fertilizer per plant.

Slugs and snails are common pests of artichokes. Try handpicking, barriers, traps, and baits. The artichoke plume moth lays eggs in the bud. Where these pests are most troublesome, a regular insecticide program may be called for.

Harvest artichokes when the flower buds are 4 inches across and before they start to open, cutting them with a knife or pruning shears and retaining a piece of stem, which is very tasty.

**Varieties** 'Green Globe' is a standard artichoke variety that produces green flower buds within 100 days.

**How to use** Artichokes should be boiled or steamed until tender (about 45 minutes) and brought to the table either hot or chilled. The leaves are pulled one at a time and usually dipped into a sauce (often melted butter if they're served hot or mayonnaise if they're served cold); then the tender inner flesh is eaten by pulling the leaf through the teeth. The rest of the leaf is discarded.

When all the leaves are removed, scrape away the fuzzy center with a spoon and enjoy the heart, considered a delicacy. The hearts may also be marinated or served in a lemon, oil, and herb sauce.

## ARUGULA
### *Eruca vesicaria sativa*

Known also as rocket, arugula is a leafy vegetable gaining in popularity as a salad green. Its leaves are deeply toothed and resemble turnip greens.

Arugula is a cool-season vegetable. Sow seeds thickly, ¼ inch deep, in early spring for a late spring crop and in late summer for a fall crop. Arugula can be grown over the winter in mild areas where temperatures do not drop below 25° F. Plantings in summer are ruined by the heat as well as the long days and short nights, which cause the plant to flower.

To harvest, snap off the outer leaves or remove the entire plant from the ground.

**How to use** Arugula has a zesty flavor; use it in a fresh salad or cook it the same way as other greens, topped with butter and seasonings. It is good accompanied with Parmesan cheese or oregano.

Arugula

Asparagus spear

## ASPARAGUS
### *Asparagus officinalis*

Asparagus has become one of the most popular home garden vegetables, and for some very good reasons. Expensive to buy at the supermarket, it is relatively easy to grow, and you can plant this perennial once and harvest early each spring for years.

Asparagus is the earliest vegetable that can be grown and harvested from the garden, and it thrives in most areas of the United States. Ideal climates are cold enough in winter to freeze the soil a few inches deep. (Asparagus does not thrive in the Deep South states of Florida, Louisiana, and Alabama.)

A plot 20 feet square or a row 50 to 60 feet long will keep a family of five or six well supplied with fresh asparagus.

**Preparing the soil** When planting asparagus you are building the foundation for many years of production; take the time to work the soil a foot or more deep, adding plenty of organic matter. As you work the soil, apply 4 to 5 pounds of 5-10-10 fertilizer per 100 square feet.

Asparagus can be grown from seed but is usually started with one-year-old crowns planted in late spring. This not only saves time but ensures vigor and productivity. Crowns are available from garden centers and seed companies. Select large, well-grown crowns that have many roots. Thinly rooted crowns are a common cause of weak plants. Roots must not be allowed to wither or dry out before planting.

Asparagus roots spread wide; dig trenches 4 to 5 feet apart and 8 inches deep. Spread some compost, manure, or other organic matter in the bottom of the trench and cover it with an inch of garden soil.

Set the crowns 18 inches apart in the row and cover with 2 inches of soil. As the new shoots come up, gradually fill in the trench. Water generously when the tops are developing.

To encourage heavy top growth and thick spears, follow a twice-a-year feeding program. Make one application before growth starts in the spring and a second as soon as the harvest is done.

No cutting should be done the first year and only a few stalks should be cut the second year. This way the plant develops copious fernlike foliage, which in turn builds up a large reserve of energy in the roots. (Note: If two-year-old roots are available in your area, they still require this 2-year establishment period.)

The third year after planting, the asparagus should give you 4 full weeks of cutting. Early in the season, shoots may require cutting only every third day, but as the weather warms and growth becomes faster, it may be necessary to harvest twice a day.

Asparagus may be blanched by mounding organic mulch over the beds as the spears are developing. To harvest, cut or snap off the spears when 6 to 8 inches high. A handy tool for cutting is an asparagus knife. Snapping—bending the spear over until it breaks—avoids injury to other shoots below ground. Break as low as possible.

**Varieties** Choose rust-resistant varieties. These include 'Mary Washington', a thick, straight, dark green variety tinged with purple at the tips; 'Waltham Washington', its improved hybrid descendant that has higher resistance to rust than its parent; and 'California 500', which will produce early, tender spears without winter frost, although it does not tolerate high summer heat.

**How to use** Young, fresh asparagus is excellent raw. Slice thinly on the diagonal for salads, or serve it on relish trays with sauces or sour cream dips. Asparagus may also be pickled.

Nothing destroys the appearance and taste of asparagus like overcooking. As Nero said, "Execute them faster than you cook asparagus." Spears should be cooked until flexible but never soft (5 to 15 minutes, depending upon size). For small portions use a steamer basket; for larger amounts tie into bunches and stand upright in a special asparagus cooker, a coffee pot, or an inverted double boiler.

## BEANS

Several varieties of beans in common use today were developed from beans grown by the American Indians. Beans originated in Central America but were well distributed in many parts of the Western Hemisphere before Columbus arrived.

'Roma II' bush beans

Bush snap bean field

Snap beans, also known as garden beans, green beans, and string beans, are grown for their immature green pod. (The wax bean, however, has a yellowish waxy pod.) Shell beans, including lima beans, are grown for their immature green seed. Dry beans are allowed to fully mature, then are collected and stored. Many beans may be used in one or more of these forms.

Beans may be termed bush or pole, depending upon growth habit. Bush beans grow 1 to 2 feet high and are usually planted in rows. Pole beans require the support of a trellis or stake. They grow more slowly than bush types but produce more beans per plant, over a longer period.

The many types of beans discussed here will vary in their heat requirements and the length of time needed to produce a crop. All except the fava bean group require warm soil to germinate and should be planted after the last frost in spring. Planting season length varies by climate. Knowing how long your growing season is and the number of days the crop needs to mature will enable you to determine how many crops are possible and the last planting date of the season.

Beans require nitrogen-fixing bacteria to be present in the soil. If your soil is lacking in the bacteria, a bean inoculant will certainly help. If you've grown beans without any problems, you probably do not need an inoculant; but if you've never grown them and want to be sure, an inoculant is advised.

Some bean seeds are treated with a fungicide that helps to prevent rot when the beans are sown during a cool, damp spring.

### Beginners' Mistakes

Here are some common mistakes that are sometimes made when growing beans.

**Planting too early** Bean seeds (except favas) will not germinate in cold soil. If you need to advance the season, sow seeds indoors in peat pots and set out plants when the garden soil warms.

**Leaving overmature pods** To get a full crop of snap beans, pick them before large seeds develop. A few old pods left on a plant will greatly reduce the set of new ones. Keep them picked in the young, succulent stage.

**Allowing soil to dry** Lack of moisture in the soil will cause the plants to produce "pollywogs"; only the first few seeds develop and the rest of the pod shrivels to a tail.

**Using insufficient fertilizer** Beans must make strong growth to be good sized before flowering. Mix a 5-10-10 fertilizer into the soil at the rate of 3 pounds per 100 square feet before planting.

**Spreading disease** To avoid the spread of disease from plant to plant, cultivate shallowly and only when leaves are free of dew or other moisture. Harvest only when plants are dry. Remove and discard all plants at the end of the growing season.

**Not rotating crops** Beans are subject to diseases that survive in the soil; therefore, growing sites should be alternated each season.

### Dry Beans

Sometimes called shell beans, dry beans are beans that have matured on the plant and are shelled before using. They are the product of a number of different plants; favas, garbanzos, horticultural beans, lima beans, soybeans, and yard-long beans can all be grown for dry beans. There are also several varieties in the same genus as common snap beans, *Phaseolus vulgaris*, that are familiar as dry beans.

Dry beans in the snap bean group are grown in the same manner as snap beans. They are harvested when the pods start to split open.

**Varieties** Recommended varieties include 'Black Turtle' ('Midnight'), small black beans on tall plants, 104 days; 'Pinto', deep pink speckled beans produced on a vining plant, 85 days; and 'Red Kidney', reddish brown kidney-shaped beans, 95 days.

**How to use** Dry beans are soaked in water until their moisture is restored, then boiled until tender. 'Black Turtle' is the bean used in black bean soup and many Caribbean and Central and South American dishes. Pinto beans are popular as Mexican refried beans. Red kidney beans are essential in most chilies.

### Favas, Broad Beans, Horse Beans, or Windsors *Vicia faba*

Not true beans at all, these are related to another legume, vetch. They grow in cool

Fava beans

Garbanzo bean plant in bloom

weather unsuitable for snap beans and will not produce in summer heat. They need 70 days of cool but frost-free weather to produce. In mild-winter areas they are planted in the fall for a spring crop. The plants grow 3 to 5 feet high and need a trellis or other support. Pods should be harvested when the seeds are half-grown to be used like snap beans.

**How to use** If harvested when seeds are pea sized or smaller, favas can be treated as snap beans. Normally, however, they're allowed to reach full growth, which requires shelling and peeling before cooking. Use any recipe for limas and serve these beans with plenty of butter. They have a sweet flavor and go particularly well with ham, pork, and chicken. Use them in soup, try them puréed, or steam them and dress with a sautéed mixture of onions, garlic, and parsley. Favas also make excellent dry beans.

### Garbanzos, Chick-peas, or Grams
### *Cicer arietinum*

Botanically, garbanzos are neither beans nor peas. This 1- to 2-foot-high bush-type plant is similar to snap beans in culture but requires a longer growing season, about 100 days. Garbanzos grow best in hot, dry climates. They produce one or two beige seeds in each puffy little pod. Pick in the green-shell stage or let them mature for dry beans.

Sow seeds when the soil is warm, and thin plants to 6 inches apart. Successive plantings will ensure a continuous supply. Harvest when the pods begin to split.

**How to use** Garbanzos can be picked green and eaten raw, but the dried, cooked form is most familiar. Whole garbanzos enhance a tossed or mixed-bean salad, may be simmered with sautéed onions and served with mushroom sauce, or can be added to soups or casseroles.

Garbanzos are a prime ingredient in the Spanish olla podrida, a hearty meat and vegetable stew. The Middle East has given us puréed beans with garlic, lemon juice, sesame paste, chopped red chiles, and spices. Vegetarian cooks often combine garbanzos with grains to form high-protein meat substitutes.

### Heirloom Beans

Hundreds of varieties of beans that were once popular have been lost to general circulation. Some have been preserved by individual gardeners or families and are called by the general term *heirloom beans.* A variable lot, they are above all else interesting and fun for the experimenter.

### Horticultural Beans
### *Phaseolus vulgaris*

These large-seeded beans are grown primarily for use in the green-shell stage, the fiber of the pod being too tough for use as snap beans. The mature pods are colorfully striped or mottled, usually with red.

Horticultural beans are grown in the same way as snap beans. Harvest when the pods are full and start to yellow.

**Varieties** 'French Horticultural' is a compact plant that does not need staking. Pods are mottled in off-white and red; beans are the same colors. Matures in 68 days.

**How to use** This bean is the French flageolet, used exclusively by the better French restaurants and served wherever French gourmets gather. To savor their superb rich, meaty flavor, they must be eaten in the green-shell stage. Cook in the manner of fresh limas. Sauté in butter, add a dash of tarragon, and serve with roast leg of lamb. Flageolets also freeze and can well, and as dry beans are prized for the way they hold their shape during cooking.

### Lima Beans
### *Phaseolus lunatus*

Limas need warmer soils than snap beans to germinate properly, and higher temperatures and a longer season to produce a crop. If days are extremely hot, however, pods may fail to set. If the soil temperature is below 65° F, pre-treating seed with both an insecticide and a fungicide before sowing is good insurance.

Lima beans may be large seeded or small seeded; the small-seeded varieties are also called baby limas or butter beans. Lima beans are also available in bush and pole types; all baby limas are bush beans. Pole beans need a longer growing season but can be harvested over a longer period of time.

Lima beans are grown in the same way as snap beans except that they need a longer

Lima beans

Scarlet runner bean flowers

and hotter growing season and should be spaced 8 inches apart. Harvest lima beans as soon as the pods are well filled but while still bright and fresh in appearance.

**Varieties** For bush limas consider 'Baby Fordhook', a small-seeded variety, 3 or 4 dark green beans per pod, 70 days; 'Fordhook', a large-seeded variety, 3 or 4 beans per pod, 75 days; 'Fordhook 242', a large-seeded variety, very heat resistant, an All-America Selection, 75 days; 'Henderson Bush,' a small-seeded variety with creamy white beans, 65 days; and 'Jackson Wonder', a small-seeded variety with beige beans marked in purple, extremely heat and drought tolerant, 65 days.

The most popular pole limas are 'Carolina' (or 'Sieva'), medium-green beans that turn white when dried, 78 days; 'King of the Garden', large, flat beans, 88 days; and 'Prizetaker', the largest pole lima bean, 90 days. In Florida or other very hot areas, look for 'Florida Butter' (or 'Calico'), beige beans splashed with purple, 90 days.

**How to use** All lima beans should be shelled before cooking. To hull limas press firmly on the pod seam with your thumb. Beans should pop out easily. Steam or simmer them for 20 to 40 minutes, depending on size. When cooked they have a mealy texture and nutty flavor.

Try limas with sautéed onions and mushrooms or crumbled bacon, or with a dressing of tomato sauce, warm sour cream, or lemon butter and dill. Lima bean soup, simmered all day with a ham bone, makes a hearty supper. Mix limas with corn for an old southwestern favorite, succotash, or serve cold with red onions and parsley in a vinaigrette dressing.

### Scarlet Runner Beans
### *Phaseolus coccineus*

These beans are closely related to common snap beans but are more vigorous and have larger seeds, pods, and flowers. The plant will grow rapidly to 10 and even 20 feet, forming a dense yet delicate-appearing vine with pods 6 to 12 inches long. Because of its large clusters of bright red flowers, it is often grown as an ornamental.

Only pole varieties are commonly available. Culture is the same as snap beans, but give them more space.

**How to use** Young pods picked when the seeds just start to develop may be cooked as green beans. Older pods are tough and strong flavored. Seeds may be shelled and are delicious when used like limas; however, though sometimes even sold as limas, they lack their distinctive flavor. Dry, the large black and red beans are strong flavored and, after cooking, somewhat astringent. They remain popular in the Southwest, where they are often used like pinto beans in chili recipes.

### Snap Beans
### *Phaseolus vulgaris*

Snap beans are said to be the foolproof vegetable for the beginning gardener. They require only about 60 days of moderate temperatures to produce a crop of green pods. With such a short growth period, they can be grown throughout the United States, and in most areas can be harvested over many months from small plantings made every 2 weeks. In long-season areas, snap beans may be grown virtually all year, but they should be planted so podding does not occur when the weather is too hot or too cold.

Snap beans are available in bush and pole varieties. Bush snap beans are slightly more cold hardy than pole types and generally can be planted as much as 2 weeks earlier. They produce beans for about 3 weeks, so they must be planted in succession for a continuous supply. Pole beans can be harvested for about 2 months. Bush snap beans are also somewhat less susceptible to heat and drought than are pole beans.

Most snap beans are green; the yellow varieties are known as wax beans. There are also several varieties with purple or purple-striped pods.

Snap bean seeds should be sown in the garden after the danger of frost has passed and the soil is warm. Sow seeds 1 inch deep in heavy soil and 1½ inches deep in sandy soil. Space bush beans 3 to 4 inches apart, except for the varieties with flat pods, which need more space. Pole beans should be spaced 6 to 10 inches apart on a trellis, or several plants to a pole.

**Varieties** The following are always near the top of the list of recommended bush snap

Purple snap beans

'Frostbeater' soybeans

beans: 'Blue Lake Bush', 6-inch pods mature simultaneously, 58 days; 'Contender', slim dark green pods, disease and adverse weather resistant, 49 days; 'Derby', an All-America Selection, disease resistant, compact plant, 60 days; 'Green Crop', an All-America Selection, long-lasting flat pods, 51 days; 'Greensleeves', dark green pods, white seeds, disease resistant, 56 days; 'Roma II', also called 'Romano', (bush form of the pole 'Romano'), flat green pods, disease resistant, 53 days; 'Royal Burgundy', round purple pods turn green when cooked, 51 days; 'Tendercrop', crisp dark green pods, 53 days; 'Tenderpod', an All-America Selection, thick, cylindrical deep green pods, 50 days; and 'Topcrop', slender medium-green pods, 49 days.

Three varieties of bush wax beans are consistently rated high: Burpee's 'Brittle Wax', high yields of 7-inch pods, 52 days; 'Cherokee', 6-inch bright yellow pods, disease resistant, 52 days; and 'Goldcrop', an All-America Selection, very disease resistant, shiny yellow pods easy to pick, 54 days.

Favored varieties of pole beans include 'Blue Lake', medium-green pods with white seeds, 60 days; 'Kentucky Wonder', used fresh or as dry beans, oval, thick, 9-inch pods, 65 days; 'Romano', sometimes called Italian pole beans, wide, flat pods, 60 days; 'Selma Star', straight, 8-inch pods, 60 days; and 'Selma Zebrina', an improvement of 'Selma Zebra', higher yielding, pods mottled in red, 58 days.

The outstanding pole wax bean is 'Burpee Golden', wide, flat yellow pods, 60 days.

**How to use** Bush and pole snap beans can be prepared in the same way: Break off both ends of the pod, then cut into 1-inch pieces, or for French-style beans, quarter lengthwise. Boil fresh, young beans in vegetable stock for 8 to 10 minutes, then serve with butter, salt and pepper, and a few drops of lemon juice. Or simmer with fried bacon squares and chopped onion; or steam briefly, then sauté in olive oil with garlic.

### Soybeans
***Glycine max***

Soybeans were cultivated in China in 3000 B.C. and since earliest times have been an all-important food in Manchuria, Korea, and Japan. First brought to the United States in 1804, they were used mainly as a forage crop until 1920. In 1942 the wartime demand for edible oils and fats created a boom for seed in commercial quantities.

An exceptionally rich source of protein and a staple of diets throughout the world, soybeans are now being recognized as a superior home garden vegetable.

Soybeans are grown similarly to snap beans. Thin seedlings to 4 inches apart. Water well, especially during dry periods. Soybeans are tolerant of high heat.

Avoid cultivation or harvest when the plants are wet and thus easily bruised and broken. Wet leaves also facilitate the spread of disease. Harvest for green beans as soon as the pods are plump and the seeds nearly full sized but still green. All the beans on the plant will ripen at about the same time, so you might as well pull the plant, then find a shady spot to pick off the pods.

**Varieties** The most widely planted variety is 'Prize', which produces 2 to 4 beans per short, hairy pod. Matures in 85 days.

**How to use** Most home gardeners use soybeans in the green stage. The pods are easily shelled if plunged into boiling, salted water for 5 minutes, allowed to cool, then gently squeezed. Shelled soybeans take about 15 minutes to cook and are good with any garnish used for limas.

The high protein content of soybeans increases the soaking and cooking time for the dried beans. A shortcut for all dried beans is to cover with water in a kettle and boil for 2 minutes, allow to stand 1 hour still covered with water, then cook until they are tender.

In dried form, soybeans can be sprouted, ground into flour, or made into soybean oil, milk, or the curd called tofu. Soybean products are often used as a meat substitute.

### Sprouts

Many different plants can be used for sprouts. The seeds of soybeans, mung beans, snap beans, lentils, peas, garbanzos, radishes, alfalfa, cress, and cabbage can all be sprouted, each having its own distinct flavor. Look for untreated seeds to use for sprouts.

Yard-long beans

Red cabbage

Sprouts mature quickly and are usually ready for harvest within a week of sowing. They can be grown in bowls, jars, or most simply by spreading seeds over wet paper towels laid in a baking pan or a shallow dish. Place the seed tray in a warm area with good light and air circulation. Cut off the green sprouts with scissors when ready to use.

**How to Use** Sprouts are used in Asian cooking, as a garnish on salads, and also as a filling for sandwiches.

**Yard-long Beans or Asparagus Beans**
***Vigna unguiculata sesquipedalis***

Although the pods of this plant resemble very large snap beans, the yard-long bean is actually a vining variety of the cowpea. Train this vigorous climber on wire or some type of trellis.

Grow yard-long beans the same way as you would cowpeas (see page 108). Harvest pods when young and before they change color, unless they are being grown for use as dried beans.

**How to use** Treat as snap beans. Break yard-longs into 1- to 2-inch pieces and cook quickly to preserve their delicate, asparaguslike flavor. Simmer 3 to 5 minutes and serve with butter and seasonings; or panfry and serve with rice. They may also be cooked as dried beans.

## BEETS

See Root Crops.

## BROCCOLI

See Cabbage Family.

## BRUSSELS SPROUTS

See Cabbage Family.

## CABBAGE FAMILY

The eight common vegetables of the cabbage or cole family—broccoli, brussels sprouts, cabbage, cauliflower, Chinese cabbage, collards, kale, and kohlrabi—are excellent home garden crops. They are grown in every climate of the United States in one season or another.

The cole vegetables are adapted to cool weather, growing best when temperatures are between 65° and 80° F. Planting should be timed for harvest also during cool weather: In cold-winter areas plant for harvest during the summer and early fall; in the South plant for harvest in late fall or winter; in mild climates plant for late spring or fall harvest.

All cole crops are frost hardy, most tolerating temperatures into the low 20s° F. (Chinese cabbage and cauliflower are the least hardy.)

All of the cole crops will grow well in reasonably fertile, moist, well-drained soils. Mulch to keep the ground cool and moist. If a soil test reveals either overly acid or alkaline conditions, raise the pH with lime or lower it with sulfur into the 6.5 to 7.5 range. (Check the label for how much to use.) A pH within this range discourages clubroot disease and permits maximum availability of soil nutrients. Rotating planting sites of cabbage and other cole crops each year also helps to avoid clubroot disease.

Work into the soil a preplanting fertilizer, such as 10-10-10, at the rate of 2½ to 4 pounds per 100 square feet. Soils known to be fertile may need only 1 to 2 pounds. Where rains are heavy or the soil is sandy, nitrogen is quickly leached. In these cases a side-dressing of about 2 pounds of straight nitrogen fertilizer per 20 feet of row will likely be necessary during the growing season.

**Planting** All the cole crops can be either direct seeded or transplanted. Direct seeding is recommended most often for the more sensitive Chinese cabbage and kohlrabi. (If you must transplant these, use peat pots or similar plantable containers.)

Space seeds about an inch apart in rows and cover with ½ to 1 inch of soil. Sow about 2 weeks earlier than you would set out transplants, to have the same maturity date.

Transplants should be 1 to 1½ months old and have 4 or 5 true leaves when set into the garden. Transplants frequently have crooked stems; these should be planted up to the first leaves to ensure a sturdy plant that will not tend to flop over when full sized.

Bolting is the rapid formation of a seed stalk, without the previous formation of the head or other harvestable product. It happens typically to the biennials: cabbage, broccoli, brussels sprouts, and kohlrabi. These will bolt if young plants with several true

Cabbage heads ready for harvest

Broccoli

leaves are exposed to temperatures below 50° F for 2 to 3 weeks. Large transplants exposed to low winter temperatures will flower in spring rather than make a crop. Bolting will be less of a problem if transplants have stems the thickness of a lead pencil.

### Broccoli
### *Brassica oleracea, Botrytis* Group

One of the most popular cole crops with home gardeners, broccoli is highly productive; unless a considerable amount is to be preserved, 6 to 12 plants will be adequate for most families. The immature flower buds are the parts of the plant that are usually eaten.

In most respects, broccoli culture is the same as that of cabbage. Most varieties will need 60 to 85 days to mature from transplants to harvest; direct-seeded plants will need another 2 weeks.

Broccoli should be planted so it is harvested when temperatures are below 80° F, so plant either 2 to 3 weeks before the last spring frost or 3 to 4 months before the first fall frost. In mild-winter areas plantings are frequently made in fall for a winter harvest.

Harvest the center green flower bud cluster while the buds are still tight and before there is any yellow showing. Except for some varieties, such as 'Premium Crop', broccoli will continue producing bonus side shoots as long as the harvested shoots are not cut back to the main stem. If you allow the base of the shoots and a couple of leaves to remain, new shoot growth will take place, and the harvest season should last a month or more. During hot weather, however, buds will pass quickly from prime condition.

**Problems** Common problems include small plants that flower early or head poorly. Bolting occurs during periods of high temperatures. Planting late in spring also contributes to this problem. Premature flowering may also be caused by extended chilling of young plants, extremely early planting, transplants that are too old or too dry, and severe drought conditions.

**Varieties** Among the best broccolis are 'Early Emerald Hybrid', 50 days, the earliest broccoli, with rich blue-green heads that produce over a long period; 'Emperor Hybrid', 64 days, a uniform and vigorous variety with large, tight heads, good climate adaptability, and good disease resistance; 'Green Comet', 55 days, an All-America Selection, which has firm dark green heads and is a good producer of side shoots after the center head is cut; and 'Premium Crop', 58 days, an All-America Selection, which has very large blue-green heads of fine texture and excellent quality.

**How to use** Broccoli is easily prepared by boiling or steaming, but it is delicate and must not be overcooked. The leaves contain the greatest concentration of nutrients, so don't discard but cook along with the florets and stems. The florets also may be sautéed.

Broccoli takes well to many garnishes, seasonings, and sauces. It may also be served raw in a salad or accompanied by a dip.

### Brussels Sprouts
### *Brassica oleracea, Gemmifera* Group

In most ways the culture of brussels sprouts is similar to that of cabbage. The most cold tolerant of the cole family, brussels sprouts is a relatively long-season plant whose cabbagelike flavor is improved and sweetened by frost.

In most areas set out transplants in May, June, or July for fall harvest. In mild-winter areas plant later for winter and spring use. Brussels sprouts can be direct seeded, but this is usually not as successful as starting with transplants. For further information, see the planting chart on pages 63 to 65.

The sprouts mature in sequence starting at the bottom of the stalk. Remove the leaves from beneath the lowest sprouts as soon as the sprouts start to touch each other to encourage heavier production.

To hasten the development of a late crop, pinch out the growing tip when the plants are 15 to 20 inches tall. This technique is particularly helpful where winter sets in early, but it may reduce the harvest somewhat. If plants are not pinched back and a longer harvest is desired, do not gather the sprouts all at once but rather as they mature. When all the sprouts have been gathered, the most tender leaves at the plant's top may be used as greens.

Brussels sprouts

'Self Blanche' cauliflower

In late fall cut the plant off at the ground, remove all the leaves, and hang it in a cool, dry cellar. You will be able to enjoy fresh brussels sprouts throughout the winter.

**Varieties** The varieties most used are 'Green Marvel Hybrid', 80 days, an extra-early, very cold hardy hybrid with mild tasting deep green sprouts; 'Jade Cross E Hybrid', 90 days, an improvement over 'Jade Cross', a vigorous, heavy-yielding variety with larger sprouts than most, which produces well in both summer and fall; 'Lunet', 100 days, often considered a standard for quality and production.

**How to use** Boil these diminutive cabbages, then dress them with butter, lemon juice, or herb vinegar. Sprinkle nutmeg on either plain sprouts or those with sauce; or garnish with raisins, slivered almonds, or chopped walnuts. Combine steamed sprouts with butter and roast chestnuts. For a salad, marinate cooked sprouts several hours in oil and vinegar dressing, then toss with cherry tomatoes.

## CABBAGE
### *Brassica oleracea, Capitata* Group

There are three types of cabbage: green leaved, with smooth green leaves; red cabbage, with purplish red leaves; and savoy cabbage, with crinkled leaves.

In the South, cabbage can be grown in all seasons except summer. Farther north, it can be planted in early spring for an early summer harvest, or in midsummer for a fall harvest.

Cabbage can be direct seeded or grown from transplants. To produce large heads, space plants 20 inches or more in rows 36 inches apart. Smaller heads can be grown spaced 12 inches apart.

Cabbage, like other cole crops, is a heavy feeder, and the best quality comes from quick growth. Before planting, add 6 to 8 pounds of 5-10-10 fertilizer per 100 square feet and work it into the soil. Follow up in 3 to 4 weeks with a side-dressing of about 1 pound of ammonium nitrate per 100 feet of row.

Cabbage responds very favorably to the cool, moist soil under a mulch.

Begin harvesting when heads are firm and glossy and about the size of a softball. Cut just beneath the head, leaving some basal leaves to support the new growth of small lateral heads.

**Problems** In warm weather the heads of early varieties tend to split soon after they mature. One solution is to plant at any one time only the number you can use during the 2- to 3-week maturation period. Another approach is to hold off on water or partially root-prune the plant when heads are formed. (Some gardeners simply twist the plant enough to break some of the roots.)

### Varieties

Those varieties resistant to yellows, a common disease of some northern states and much of the upper South, are indicated by YR.

**Green-leaved varieties** 'Copenhagen Market', 72 days, vigorous with round, medium-sized heads; 'Danish Ballhead', 105 days, with solid, medium-sized heads that are good for storing; 'Early Flat Dutch', 85 days, with large, flattened, medium green heads that are split resistant; 'Early Jersey Wakefield', 63 days (YR), a variety with small to medium conical heads with a sweet, mild flavor; 'Emerald Cross Hybrid', 63 days, an All-America Selection, a sweet-tasting cabbage with blue-green leaves and round, solid, medium-sized heads; 'Golden Acre', 64 days (YR), a round, firm, medium-green headed variety; 'Golden Cross', 40 days, a good hybrid variety for small gardens that has tight, sweet, golden yellow heads; 'Late Flat Dutch', 100 days, a vigorous variety with large, flat, medium-green heads; 'Rio Verde', 87 days, with large, slightly flat heads, very popular in the South and also fairly cold tolerant; and 'Stonehead', 70 days (YR), an All-America Selection, a hybrid with extremely solid heads, blue-green leaves, and good holding quality in the ground.

**Red varieties** Among the most popular are 'Red Acre', 76 days, with solid deep red heads on fairly compact plants; 'Ruby Ball Hybrid', 68 days, an All-America Selection, a hybrid with small round red heads that hold well at maturity, and 'Ruby Perfection Hybrid', 85 days, a uniform, solid, globe-shaped cabbage that is slow to split.

Cauliflower with leaves tied

Chinese cabbage

**Savoy varieties** These have become deservedly more popular with many home gardeners. Recommended are 'Savoy Ace Hybrid', 78 days (YR), an All-America Selection, a semiglobular, firm cabbage that grows on a medium to large plant; and 'Savoy King Hybrid', 90 days (YR), a high-yielding cabbage with semiglobed deep green heads.

**How to use** The strong cooking odor of cabbage can be cut by simply tossing a stalk of celery into the pot. Enjoy cooked cabbage with corned beef, and in Spanish and Italian boiled dinners. Stuff the head with ground pork or veal, or make cabbage rolls of leaves filled with ground meat or rice and cheese.

Pickled cabbage, or sauerkraut, is an international favorite—with pork and sausages in the French *choucroute garnie*; in the Russian sauerkraut soup, or on a plain old U.S. hotdog.

### Cauliflower
### *Brassica oleracea,* *Botrytis* Group

More restricted by climate than either cabbage or broccoli, cauliflower is less cold tolerant and will not head properly in hot weather. Its general culture and season length, however, are about the same as those for early- to medium-season cabbage. The edible part of cauliflower is the head of tightly clustered white or purple flower buds that are surrounded by large blue-green leaves.

The approximately 2 months that cauliflower transplants need to mature must be cool. This means planting for spring and fall crops in most cases, although winter crops are possible in mild-winter areas and summer crops in some gardens at high elevations. For a spring crop, plant transplants a week or two before the average date of the last frost.

For the most vigorous seedling growth, cauliflower is nearly always grown as a transplant. Transplants must grow rapidly; if old or stunted they will produce buttons—very small heads atop immature plants. Start seeds 1 to 2 months before the outdoor planting date.

As plants mature in the garden and the curd (head) begins to develop, gather the leaves over it and tie them together with soft twine or plastic tape. This is called blanching, because with light excluded the curd will turn from yellow to white and become tender. The leaves of the variety 'Self Blanche' curl naturally over the head when grown in cool weather. Harvest as soon as the heads fill out by cutting the stalk below the head.

If the weather turns hot, mist or sprinkle the plants to maintain humidity and cool temperatures. Unwrap the heads occasionally to check for hiding pests.

**Varieties** The most widely grown varieties include 'Early Snowball', 60 days, medium-sized plants and smooth white heads; 'Early White Hybrid', 52 days, a plant with very large, round white heads and leaves that tend to self-blanch the heads; and 'Purple Head', 85 days, with heads that are deep purple on top and turn green when cooked. This variety does not need blanching and holds well in the garden. Others widely grown are 'Snow Crown', 50 days, an All-America Selection, a hybrid that produces deep, rounded pure white heads even under somewhat adverse conditions; and 'Snow King', 45 days, a hybrid grown where extra earliness is needed.

**How to use** Raw cauliflower is delicious dipped into *bagna cauda,* an Italian anchovy-garlic-butter sauce, and is also good with most popular American dips.

To serve cauliflower hot, steam and cover with a nutmeg-spiced cheese sauce in the Scandinavian manner, or bake the parboiled florets with melted butter and grated sharp cheese. Cooked cauliflower goes well in many quiches and casseroles.

### Chinese Cabbage or Celery Cabbage
### *Brassica rapa,* *Pekinensis* Group

Chinese cabbage is more closely related to mustard than to cabbage. Although it is unrelated to celery, it is sometimes called celery cabbage, because the loose egg- or vase-shaped plants have the same shape as celery plants. Chinese cabbage is a cool-weather crop and often bolts in the long days of late spring and summer. It is grown most successfully as a fall and early-winter crop.

Collards

Kale (foreground)

Sow seeds thinly in rows 24 to 30 inches apart, then thin seedlings to 18 inches apart. If you must transplant, start seeds in a peat pot or similar plantable container.

Plant from early August into September. Chinese cabbage requires 75 to 85 days from seed to harvest. Pull up the entire plant when it is mature and strip away the outer leaves. If frost hits before the heads form, you will still have a good crop of greens.

**Varieties** 'Michihili Jade Pagoda Hybrid', 60 days, is a thick, cylindrical plant with a crinkled cream-colored center. 'Two Seasons Hybrid' has large, oval heads that hold up well in warm weather and resist bolting. It can therefore be grown as a spring crop as well as in the fall.

**How to use** This vegetable is traditional in Asian soups, sukiyakis, and stir-fried dishes and can be used raw in salads. Butter-steam the leaves to accompany duck; for cabbage rolls, stuff with a mixture of minced chicken and pork. For slaw, shred fresh Chinese cabbage and add fresh pineapple or carrots.

### Collards
### *Brassica oleracea, Acephala* Group

Like kale, this perennial is one of the oldest members of the cabbage family. Europeans described it during the first, third, and fourth centuries, and Colonial American gardeners wrote about it in 1669.

Unlike their close relative kale, collards withstand considerable heat, yet they still tolerate cold better than cabbage. Collards do not form a true head like cabbage but instead grow in a large rosette of blue-green leaves.

Planting in spring and again in fall will produce a supply of greens almost every month in the year in all but the coldest areas of the South. (A light freeze sweetens the flavor.)

Three different planting methods can be used with collards: In spring, sow seed or set out plants to stand 10 to 15 inches apart; plant close, 5 to 7 inches apart, to dwarf and make bunchy plants for harvesting leaves as needed; or in summer, sow seed thinly and let seedlings grow until large enough for greens, then harvest seedlings to give 10- to 15-inch spacing.

Collards have the same fertilizer and water requirements as cabbage.

**Harvesting** Successive plantings are not necessary for a continuous supply. Collards are one of the most productive of all vegetables, particularly in southern gardens. Harvest seedlings or entire plants, or gradually pick lower leaves.

**Varieties** Best are 'Georgia', 70 to 80 days, a vigorous plant that produces tender, juicy greens; 'Hicrop', 75 days, the first hybrid collard with a mild, sweet flavor and good texture, slow to bolt in hot weather; and 'Vates', 75 days, a low-growing, compact variety with thick, broad leaves.

**How to use** Collard greens are most often served boiled with salt pork or hog jowls. The resulting juice is called pot likker and is eaten with hot corn bread. Collards also traditionally accompany pan-fried fish and are common in southern-style pea soup. Leaves harvested in warm weather will have an improved flavor if they are refrigerated for several days before use.

For variety, cook and drain collards, top with grated sharp cheese, and bake until the cheese is bubbly. Or make a hot, wilted salad by pouring bacon drippings over shredded young collards mixed with chopped green onions.

### Kale
### *Brassica oleracea, Acephala* Group

Kale is the closest relative of the wild cabbage from which all of the cole crops have been derived; it has been raised as long as man has raised any vegetable. Grown for its finely cut or curled blue-green leaves, which are produced in rosettes, kale has adapted better than any other vegetable to fall sowing throughout a wide area of the United States, including areas of moderately severe winters. However, although kale is extremely hardy, it does not tolerate heat quite as well as collards.

Kale grows best in the cool of fall, and the flavor is improved by a frost. Transplants may be used for early spring planting, but direct seeding is best in fall. Since kale thinnings are good eating, scatter seeds in a 4-inch band in rows 18 to 24 inches apart, then thin seedlings to 8 to 12 inches. In other respects, their

Kohlrabi

Cardoon

culture is essentially the same as that of cabbage.

Harvest kale either by cutting down the entire plant or by removing the outer leaves.

**Varieties** 'Dwarf Blue Curled Scotch Vates', 55 days, is the most widely planted kale. It is a low, compact, spreading plant with finely curled leaves.

**How to use** Always remove kale's tough stems and midribs before cooking. Chop the leaves, then proceed with any cooking method you would use for spinach.

### Kohlrabi
### *Brassica oleracea, Gongylodes* Group

A relatively recent development from wild cabbage in northern Europe, kohlrabi was not known 500 years ago and was not noted in the United States until about 1800.

This unusual, little-known vegetable deserves to be grown and appreciated more. It looks like a white or purple turnip growing above ground and sprouting long stems and leaves. Kohlrabi tastes like cabbage, but is sweeter.

In most regards, kohlrabi is similar to cabbage in culture, but with a shorter season, it should be started in early spring so that most growth is complete before the full heat of summer. (Kohlrabi is, however, more tolerant of heat and drying winds than other cole crops, and it is one of the few vegetables that will withstand light shade.) Start in midsummer to harvest in fall.

Kohlrabi is usually seeded directly into the garden but can be grown from transplants. Space plants about 4 to 6 inches apart in rows 1 to 2 feet apart. Harvest while the plants are 2 to 3 inches in diameter.

**Varieties** Recommended kohlrabi varieties include 'Early Purple Vienna', 60 days, which is later and larger than 'Early White Vienna' and has greenish white flesh that is purple on the outside; 'Early White Vienna', 55 days, with uniform 2-inch bulbs; 'Grand Duke', 45 days, an All-America Selection that is a large, compact, early bearing plant with crisp, mild white flesh. Its leaves may also be eaten.

**How to use** Use the enlarged stem (bulb), discarding the leaves. Young kohlrabi makes a delicious chilled salad or can be eaten out of hand like an apple. Or cut it into strips for a relish tray. Try marinating it in a dressing of mayonnaise and sour cream seasoned with mustard, dill seed, and lemon juice.

When small and tender, kohlrabi is good steamed, without peeling. (As it matures it's best to strip off the tough, fibrous skin.) Dice or quarter, boil in a small amount of water, then serve with a creamy cheese sauce seasoned lightly with nutmeg.

### Pak Choi
### *Brassica rapa, Chinensis* Group

This close relative of Chinese cabbage is grown for its stalks, which are used in chow mein and stir-fried dishes or eaten raw. Its deep green leaves cluster but do not form a head.

Favoring cool weather, pak choi can be grown either as a spring or fall crop, or as a winter crop in the South. Otherwise, it is grown in the same way as Chinese cabbage.

**Varieties** 'Lei-Choi', 47 days, has spicy, celerylike pure white stalks; 'Mei Qing Choi', 45 days, has sweet-flavored pale green stalks and is smaller than most other pak choi varieties.

## CABBAGE

See Cabbage Family.

## CANTALOUPES OR MUSKMELONS

See Melons.

## CARDOON
## *Cynara cardunculus*

This handsome plant, with its silvery fernlike foliage, is a favorite with Italian and French cooks. Because of its ease in growing and preparation, it deserves more recognition in American cuisine.

Closely related to the artichoke, cardoon has a flavor somewhat between celery and zucchini. Physically it resembles the artichoke, with ornamental 3- to 4-inch thistlelike purple flowers, deeply cut leaves, and a crown that multiplies by sending out side branches. But whereas the artichoke is raised for its fleshy flower bud, cardoon is grown for young leafstalks and occasionally for its root.

Plant cardoon in the garden where it can be appreciated close up. Start seeds indoors 10 weeks before the last

'Tall Utah' celery

Celeriac

frost, then move the plants to the garden and space them 1 to 2 feet apart. A cool-season plant, cardoon requires 120 to 150 days from seed to harvest. It needs rich, moist soil; keep plants well fed and watered for vigorous growth. If the plants have to struggle to survive, the leafstalks will become pithy and the plant will put its energy into flowering.

Cardoon is a perennial where winter temperatures do not drop below 20° F. Elsewhere, it is usually grown as an annual.

When plants are 3 feet high, blanch the stems to improve their flavor and quality. Do this by tying them together and wrapping them in paper or burlap. Harvest by cutting off the blanched stems just below the crown, and trimming off the outside leaves.

**How to use** Cut cardoon stems into sections and parboil in salted water and lemon juice (to prevent darkening) until tender (an hour or more). Serve as a chilled salad with a vinaigrette dressing, or as a hot vegetable seasoned with butter, cheese, or a light cream sauce. Italian cooks prefer it dipped into a light egg batter and deep-fried until just crisp.

## CARROTS

See Root Crops.

## CASABA MELONS

See Melons.

## CAULIFLOWER

See Cabbage Family.

## CELERY AND CELERIAC

These members of the parsley family probably originated near the Mediterranean, but wild forms barely resembling cultivated varieties grow in low-lying wet places throughout Europe and southern Asia. The French apparently were first to use celery as a food, around 1600. Earlier it found some use as a medicine. The first recorded commercial production of celery in the United States was in Kalamazoo, Michigan, in 1874.

**Celery**
***Apium graveolens* var. *dulce***

Celery is a leafy plant that produces long, edible, crisp stalks. It demands more time and attention than most garden vegetables, and needs about 4 months of temperatures over 70° F. If you don't start from transplants, sow seed indoors 2 to 4 months before spring planting time. Seed should germinate in 2 to 3 weeks.

Sow the very small seed ⅛ inch deep in flats or pots and keep moist by covering with damp burlap. Transplant seedlings carefully after the frost danger has passed; provide generous shade and moisture to the new plants until they are established.

Celery grows naturally in wet, almost boggy locations, so the water supply must be plentiful and continuous.

Use plenty of 5-10-10 fertilizer, since celery is a long, heavy feeder. Soil should be rich in organic matter and have a neutral pH.

Although blanching is not usually necessary with modern varieties, it may make them more tender. White stalks are becoming less common in markets, so you may want to grow your own. Wrapping with paper, shading with boards, or mounding soil around the stalks will blanch them. Celery is harvested by digging the whole plant from the ground or by pulling off the outer stalks, leaving the inner stalks to grow.

**Varieties** Use slow-bolting varieties for early spring planting: 'Golden Self-Blanching', 115 days, with long, thick-ribbed, stringless golden yellow stalks, and tolerant of cold spring temperatures; or 'Utah 52-70 Improved', 105 days, a disease-resistant variety with extralong dark green stalks.

For late spring or summer planting, use 'Fordhook', 130 days, a stocky, compact plant whose crisp, juicy stalks store well; or 'Giant Pascal', 125 days, a long-stalked celery with creamy white stalks.

**How to use** Celery is well known as an hors d'oeuvre stuffed with cheese or served with hot or cold dips. Add to salads or use chopped to enhance spreads. The leafy tops, chopped fine, go well in soups and salads and can also be

Celtuce

'Rhubarb Chard'

dried, powdered, and used as a seasoning.

Celery makes a delicious creamed soup. As a hot vegetable it can be boiled, braised, fried, or baked. Try stewing with tomatoes, shallots, and basil; or serve hot garnished with anchovy fillets and sprinkled with wine vinegar.

**Celeriac**
***Apium graveolens* var. *dulce***

A form of celery grown for its swollen, rough, globular root, this plant is smaller than celery and its foliage a very dark green. It is sometimes called turnip-rooted celery.

Grow celeriac in the same way as celery; it is just as demanding of plentiful fertilizer; rich, neutral soil; and a continuous supply of water. Any side shoots that develop should be removed.

Harvest the root once it is 2 inches or more in diameter, discarding the top. Roots may be mulched in fall to extend the harvest period.

**Varieties** 'Alabaster', 120 days, is the most common variety, and stores well.

**How to use** The flavor of celeriac has been described as a combination of celery and English walnuts. Although celeriac can be shredded and served raw in salads, it is better cooked, and is good in soups, stews, and Asian dishes. Try it steamed and served with butter or a cream sauce; or parboil, slice, and bread it, then fry in butter.

The faint bitterness of celeriac can be removed by blanching in salted water and lemon juice just prior to preparation.

## CELTUCE
## *Lactuca sativa* var. *asparagina*

Native to China, celtuce was named because it combines the uses of lettuce and celery. It is sometimes called asparagus lettuce for the same reason. The leaves, when young, are harvested for fresh salad or cooked greens. When the plant matures, the thick, central core can be used like celery or asparagus.

Celtuce is grown in the same manner as lettuce. It grows well during the winter in frost-free areas. Greens are ready for harvest in 45 to 50 days; the central stalk is ready in 90 days.

**How to use** Celtuce leaves are used in salads in the same way as lettuce, or cooked as greens. The stalk, once it is peeled, can be eaten either raw or cooked. Raw celtuce is very high in vitamin C (cooking destroys most of this vitamin). Its consistency when cooked is similar to that of artichoke hearts.

## CHARD
## *Beta vulgaris*, *Cicla* Group

Chard, or Swiss chard, is a kind of beet whose edible parts are leaves and stalks instead of roots. Considered the beet of the ancients, it was popular in antiquity, long before Roman times.

Chard's greatest virtue is its ability to take high summer temperatures in stride whereas other greens such as spinach and lettuce bolt. Chard tolerates poor soil but grows best in open, rich, neutral, well-drained soil.

Plant chard at the same time as beets: fall to early spring in mild-climate areas, and spring to midsummer in cold climates. Sow seeds in rows 18 to 24 inches apart, and thin seedlings to 4 to 8 inches apart. Thinnings can be used for greens. Chard does not grow successfully from transplants.

Because the large, outer crinkly leaves and fleshy stalks can be cut as the plant grows, one planting can be harvested over many months. Even if the entire plant is cut off an inch or two above the crown, new leaves will come.

**Varieties** 'Fordhook Giant', 60 days, develops very broad, thick white stalks and thick, crinkly dark green leaves with white veins.

'Lucullus', 60 days, has crisp, curled light green leaves and broad white stalks.

'Rhubarb Chard', 60 days, has textured, dark green, red-veined leaves and red stalks.

**How to use** Chard stalks can be cooked like celery and its leaves like spinach; if stalks and leaves are cooked together, the stalks should be given a 5-minute head start for equal tenderness.

Cut the thick stalks into 2- or 3-inch lengths and simmer in boiling salted water until

Chayote

Witloof chicory

tender. Serve hot with butter and a touch of wine vinegar, or chilled with a vinaigrette. A popular Italian dish combines cooked chard topped with butter and Parmesan cheese.

## CHAYOTE
## *Sechium edule*

Chayote, also known as vegetable pear and mirliton, is a member of the gourd family but certainly doesn't look like a gourd. In mild-winter areas it grows as a perennial. Frost will kill back the tops but the vine renews itself in spring if the ground does not freeze. Chayote is fast growing and best supported by a trellis or fence. Flowers appear in late summer and fruit is ready for harvest about a month later, continuing until frost.

Plant chayote in the spring after all frost danger has passed. The whole fruit is used as the seed and can be obtained from the market. Place the fruit on a slant with the wide end down, stem end slightly exposed. The vines are vigorous and grow quite large; one plant can produce 3 dozen or more fruits, more than enough for most home gardens. Vines need at least 6 months of warm, frost-free weather to bear a usable crop.

In cold climates mulch the roots heavily with compost or similar material for winter protection. Pull the mulch aside in spring at sprouting time. The growing plant will need plenty of water and fertilizer but not too much nitrogen, which will produce excessive vine growth.

Pick chayote when it is 4 to 6 inches long. Store chayote fruits in a cool place; they will keep for 2 to 3 months for later eating or for seeding in the spring. If the plant sends out shoots in storage, which is likely, cut them back to 2 inches when you plant.

**How to use** Chayote can be used in more ways than squash. It may be diced and steamed until tender, baked and stuffed, cooked and marinated for use in cold salads, or pickled and candied. A favorite food of Mexican cooks, it takes seasonings well and complements many dishes.

Young chayote can be cooked without peeling. Large, fully mature fruit will have a tough skin. Cut into slices, right through the flat inner seed. It has a nutlike flavor after cooking.

## CHICORY
## *Cichorium intybus*

Related to endive, escarole, and radicchio, chicory is a warm-season vegetable. The names chicory and endive are sometimes used interchangeably, although this is not technically correct. Chicory has oblong, basal foliage, which forms a loose head of dark green leaves that can be used in salads. The roots of 'Magdeburg' chicory are ground and used as a coffee substitute or additive. What is known as French or Belgian endive is actually a chicory, called witloof chicory, which is forced and blanched in the dark to produce tubular heads of leaves, usually but not always indoors in winter.

Chicory seeds are planted outdoors after all danger of frost has passed, ½ inch deep and 4 inches apart in rows 18 inches apart. Thin plants to 12 inches apart as they grow. Chicory grown for greens needs average soil; chicory grown for roots requires a soil that is loose, rich, and moist. Fertilize at planting time and again in 2 months.

To harvest chicory for leaves, pick them at any time until frost, when they become too bitter. Plants are hardy to -40° F and can be left in the ground all winter, and new growth harvested in spring. If allowed to grow as a perennial, chicory produces blue flowers in summer. For roots, harvest after the first fall frost.

To grow French or Belgian endive, grow witloof chicory from early summer to fall. Then, after cutting back the tops and digging up the plant, set the roots in moist soil in a warm cellar and cover with a 6-inch layer of moist sand. New leaves will grow in the sand and produce tight, blanched heads.

In mild-winter areas, the same results can be achieved in the garden. Sow seeds in early summer, not before, because a plant that goes to seed is useless for forcing. Thin seedlings to 4 to 6 inches apart. When plants are fully grown and the first frost hits,

Chicory plant in bloom

Corn

cut off the tops 2 inches above the crown to prevent injury to the crown buds, then cover plants with 6 to 8 inches of soil. In late December start harvesting in the first few feet of the row by removing the soil to expose the white shoots. After cutting the shoots, scrape the soil back over the roots in a mound to force a second crop. Drive a stake to mark how much row you have harvested. The harvest can continue through winter and early spring.

**Varieties** Some favorite varieties are 'Large-Rooted Magdeburg', about 15 inches tall, upright dandelion-like leaves, root 12 to 14 inches long, young tender leaves can be harvested for greens at 65 days, roots mature at 120 days; 'Sugarhat', grown for sweet yet tangy leaves, resembles romaine lettuce, 86 days; and Witloof, the chicory used to produce Belgian endive.

**How to use** The leaves of chicory are used alone or mixed with other greens in salads. They have a slightly bitter flavor. Chicory roots used for coffee should be dried and ground immediately after harvest. Add an ounce of ground chicory to a cup of ground coffee before brewing for a coffee reminiscent of that served on the Left Bank of Paris or in New Orleans. French or Belgian endive (the forced greens are called *chicons*) is used fresh in salads; a winter favorite in France is a combination of chicons and beets. Endive can also be steamed until tender (about 10 minutes) and topped with a seafood creole.

## CHINESE CABBAGE OR CELERY CABBAGE

See Cabbage Family.

## CHIVES

See Onion Family.

## COLLARDS

See Cabbage Family.

## CORN
## *Zea mays* var. *rugosa*

The word *corn* has had many meanings. Originally it meant any hard particle of grain, sand, or salt. Corned beef earned its name because it was cured with salt. Both wheat and barley were called corn in the Old World. Maize, the main cereal of the New World, was first known as Indian corn and later just corn.

Corn supported the early civilizations of the Americas. Fossils show that corn was grown in North America more than 4,000 years ago. It was after the discovery of America that corn spread rapidly throughout the Old World.

Dent corn is the corn most widely grown by farmers. It's not sweet at maturity and has an indentation, or dent, atop each kernel. It is usually grown to feed farm animals.

Most home gardeners grow sweet corn which differs from other types of corn by its ability to produce and retain more sugar in the kernels. This characteristic is controlled by a single recessive gene called sugary-1. Other distinguishing characteristics of sweet corn include tender kernels at edible maturity, refined flavors, a tendency to produce suckers at the base of the plant, and wrinkled seeds when dried. Standard sweet corn tastes best when cooked immediately after picking and remains sweet for only a few days.

In recent years a new sweet corn called supersweet has become available. Even sweeter than standard sweet corn, it retains its sweet flavor longer, for up to 14 days, because the sugar is slow to convert to starch.

Popcorn resembles sweet corn, the main difference being that the kernels are pointed and explode when subjected to heat. Popcorn is grown in the same manner as sweet corn.

For a continuous supply of sweet corn through summer and into the fall, plant early, midseason, and late varieties, or make successive plantings every 2 to 3 weeks.

Keep in mind that the number of days to maturity is a relative figure, varying by the total amount of heat the corn receives. Corn does not really start growing until the weather warms, and it grows best where summers are long and hot. Varieties listed as 65 days may take 80 to 90 days

Tassels on corn

Corn stalks

when planted early but may come close to the 65 days if they are planted at the specified time.

When planting for a succession of harvests, the effect of a cool spring should be considered. Rather than plant every 2 weeks, make the second planting when the first planting is knee-high.

Corn does not transplant well. It should be grown from seed, sown outdoors after all danger of frost has passed and the soil temperature is 50° F or higher. Plant seeds 2 inches deep, 4 to 6 seeds per foot, in rows 30 to 36 inches apart. Thin to 10 to 14 inches between plants. You can crowd back to 12 inches or even the 10-inch minimum; but, except for the small-growing early varieties, which can be spaced 8 inches apart in rows 30 inches apart, planting any closer lowers the yield and risks a crop of nubbins.

More home garden corn plantings are ruined by overcrowding than any other factor, and too many seedlings in a row act just like weeds. If you overspace corn you usually are compensated by more usable ears and some sucker production.

Seeds of supersweet varieties need warm soil and twice the moisture of standard sweet corn to germinate.

How much fertilizer and when? At planting time, fertilize in bands on both sides of the row of seed, 2 inches from the seed in the furrow and an inch deeper than seed level. Use 3 pounds of 5-10-10 (in each band) per 100 feet of row. When the corn reaches 8 inches high, side-dress with the same amount. Repeat again when the corn reaches 18 inches.

**Watering** Corn needs continual watering, from planting until harvest. The water need is greatest from tasseling time to picking time, when sweet corn makes very rapid growth; no check in watering should occur. In very hot and dry weather, rolling of the leaves (edges turning downward) may occur in midday even when soil moisture is adequate, if the plants transpire water faster than the roots can absorb it. But if leaves roll other than at midday, check the soil for adequate moisture.

Corn is shallow rooted, so when weeding be careful not to damage the roots. Plants can be mulched, or pumpkins and squash can be interplanted with the corn, in the Indian tradition.

**Harvesting** As a general guide, corn is ready to harvest 3 weeks after the silk first appears, although this will vary depending upon the weather during that time. The silk will become dry and brown when the ears are near perfect ripeness. Probably the only sure way to tell is to open the husk of a likely ear and press a kernel. If it spurts milky juice, it is at the peak of ripeness.

Varieties differ, but most will produce two ears per plant. The top ear usually ripens a day or two ahead of the lower one.

Harvest by breaking the ear from the stalk: Hold the ear at its base and bend downward, twisting at the same time. The idea is to break off the ear close to its base without damaging either it or the main stalk.

Popcorn should be harvested at the end of the growing season when the stalks have turned brown. It should be dried in a warm area before use.

**Do's and Don'ts** Corn is wind pollinated. To ensure pollination plant in short blocks of 3 or 4 rows rather than a single long row of the same number of plants.

Don't interplant different types of corn. Pollen from dent corn will make sweet corn kernels starchy and less sweet. Popcorn will cross-pollinate with sweet corn unless it is more than 100 feet away. Yellow hybrid corn must be planted downwind from white corn or the white corn will not develop. Standard sweet varieties will reduce the quality of the supersweet kinds. Grow the supersweet varieties at least 300 to 400 yards away from standard varieties. One way to avoid cross-pollination between different types of corn is to wait about 4 weeks between plantings to ensure that the plants will not be pollinating at the same time; another way is to plant them in separate areas.

Don't worry about suckers. They don't take any strength

Assorted corn varieties

'Honey 'N' Pearl' corn

from the main stalk, and removing them may actually reduce yield.

Do look out for the corn earworm. Its eggs hatch on the silk of the developing ear and larvae burrow into and feed on kernels. Ears with tight husks and good tip covers are somewhat more resistant to corn earworm damage. They do not prevent the entrance of the worms but the damage will likely be lessened. Frequently, some control measure for corn earworms is necessary. See page 34.

Tight husks have an advantage over another pest—birds—which eat the kernels on the tip of the ear. The damage is worse on ears with a loose husk. Such damage to the ornamental Indian corn makes it worthless. One way to solve the problem is to slip a paper bag over each ear after it has been pollinated.

### Corn Varieties

Since there are hundreds of corn varieties, some confusion over what to plant is difficult to avoid, especially if you're just getting started. There are white, yellow, and bicolor varieties in both standard and supersweet types of corn. Variety names often indicate the type. They vary, of course, in the time needed to mature. Hybrids are so indicated in the variety description.

**White corn** 'How Sweet It Is' is a supersweet hybrid that takes 80 days to mature. Cobs grow 8 inches long on 7-foot plants and have 16 to 18 rows of kernels. An All-America Selection.

'Silver Queen' is a standard and by far the most widely favored hybrid white corn. Very sweet (almost too sweet for some), with great flavor and tenderness, it has 8- to 9-inch ears and 14 to 16 rows of kernels, and needs 92 days to mature. Plants grow to 8 feet and are resistant to bacterial wilt, a serious corn disease.

'Trucker's Favorite' is the all-time favorite roasting white corn, 9-inch ears with 14 rows of kernels, stalks 7 feet high, 85 days.

**Yellow corn** These are the widely available, long time favorites and new varieties:

'Earliking' has yellow kernels in 10 to 12 rows on 7- to 7½-inch ears, 66 days. It is a good variety for short summers and may yield 3 ears per plant. Plants grow up to 5 feet tall.

'Early Sunglow' is a very fast maturing yellow hybrid variety, small plants about 4½ feet tall producing 6-inch ears, 12 rows of kernels per ear, 62 days. It is a good variety for areas with cool spring weather.

'Early Xtra-Sweet' is a supersweet hybrid that matures in 70 days. Plants are 5 to 6 feet tall; cobs are 7 to 9 inches long and have 12 to 16 rows of kernels.

'Golden Beauty' has yellow kernels in 12- to 14-inch rows on 7-inch ears, 73 days, an All-America Selection hybrid that is 5 feet tall.

'Golden Cross Bantam' is the standard hybrid by which all yellow sweet corns have been measured. It is still a favored main-crop variety, with 10 to 14 rows of kernels on 7½- to 8-inch ears, plants 6 to 7 feet tall, for fresh use, canning, and freezing, 85 days.

'Golden Queen' has similar characteristics and qualities to 'Silver Queen' but has yellow kernels.

'Illini Xtra Sweet' is a supersweet hybrid that matures in 83 days. Plants are 6 feet tall and have 8-inch ears with 14 to 18 rows of kernels.

'Iochief' is an All-America Selection hybrid that is one of the most wind- and drought-tolerant corns. It has 14 to 18 rows of kernels on 9-inch ears, usually 2 ears per 6½-foot stalk, 89 days.

'Jubilee', 81 days, is a good, very sweet hybrid variety for fresh use, freezing, and canning that grows well in cooler climates. Plants are 7 feet tall and have 7- to 9-inch ears.

'Kandy Korn' is a yellow extra sweet hybrid corn identified by faint red stripes and dark red tips on its husks. It is very tender and keeps both flavor and tenderness for an extended time. It carries the EH (everlasting heritage) factor so, unlike the supersweet hybrids, it does not require isolation from other sweet corns. It has 8-inch ears with 16 to 20 rows of kernels, 8-foot plants, 85 days.

'Merit' has bright yellow kernels and is very productive, 16 to 20 rows of kernels on 8- to 9-inch ears, 5-foot plants, 84 days to maturity.

Watercress

Cucumbers grown on a trellis

'Polar Vee', 53 days, is a good variety for areas with short, cool summers. Ears are 4 to 6 inches long with 12 rows of kernels. Hybrid.

'Seneca Chief', a standard late or main-season yellow hybrid, has 12 to 16 rows of kernels on slender 8-inch ears. Sweet and tender, it is one of the most popular for home gardens. Six feet tall, 82 days.

'Super Sweet' is a supersweet hybrid with very thick husks covering 7- to 8-inch ears having 16 to 18 rows of very tender kernels. Plants grow 6 feet tall.

**Bicolor corn** Kernels are yellow and white. An increasing number of this type is becoming available.

'Butter and Sugar' has 12 to 14 rows of kernels on 7½-inch ears, hybrid plants 5 to 6 feet tall, 70 days.

'Honey and Cream' has tight 7-inch ears with 12 to 14 rows of kernels, for fresh use, roasting, and canning, 78 days. Plants 5 to 6 feet tall. Hybrid.

'Honey 'N' Pearl' is a supersweet All-America Selection, 78 days, with 16 rows of kernels on 8-inch ears. Plants are 6 feet tall.

**Popcorn** 'Creme-Puff', 105 days, is a yellow hybrid corn with 2 ears per 8-foot plant.

'Peppy Hybrid' has 2 or 3 ears per 5- to 6-foot plant; 4-inch ears have hull-less white kernels. 90 days. The fastest growing popcorn.

**How to use** Nothing equals the flavor of a fresh ear of corn cooked only until heated through—and fresh means brought directly from the garden to the cooking pot. The sooner the ears are used, the sweeter they'll be: Even 24 hours after picking will produce a substantial loss in flavor and texture (except for the supersweet hybrids). If you must store corn, wrap it unhusked in damp paper towels and place it in the refrigerator. Then husk just before cooking, using a stiff vegetable brush to help remove the silk.

Corn on the cob is usually steamed or boiled. If boiling it add a tablespoon of sugar to the water to bring out the corn's natural sweetness; never add salt, which toughens the kernels. Cook until just tender (3 to 5 minutes), then serve with plenty of butter, salt, and pepper. Ears are also delicious roasted either in the oven or over hot coals. Husk them, coat with butter, wrap in aluminum foil, and roast about 40 minutes in the oven, 15 minutes on the grill. To give this old favorite a different twist, serve the ears with melted butter seasoned with soy sauce, mixed herbs, curry powder, Worcestershire sauce, or chopped chives.

For another succulent treat, steam fresh kernels in milk or light cream. Or make corn chowder, scalloped corn, succotash (lima beans and corn), or a corn soufflé. Corn fritters with hot syrup and sausages make a delicious breakfast.

## COWPEAS OR BLACK-EYED PEAS

See Peas.

## CRENSHAW MELONS

See Melons.

## CRESS

Three kinds of cress are cultivated and more are found growing wild.

### Early Winter Cress
### *Barbarea verna*

Also called Belle Isle cress and upland cress, this hardy biennial survives severe winters. Sow it in late summer and use it in fall and winter, or allow it to remain in the ground all winter and use the following spring and summer before the seed stalks develop. It can also be sown in early spring and will be ready for harvest in 7 weeks. Sow seeds ¼ inch deep in rows 12 to 14 inches apart; thin plants to 4 to 8 inches apart. Similar in appearance but more bitter than watercress, winter cress should be cooked with another vegetable, such as spinach, to cut its strong flavor.

### Garden Cress
### *Lepidium sativum*

Also known as peppergrass and curly cress, this fast-growing annual looks like parsley. Don't cover the seed. It germinates in a few days if the

Armenian cucumbers

Cucumber blossom

seeds are exposed to light, and it can be eaten in 2 weeks as sprouts, or later in all stages up to maturity. It is a cool-season, short-day plant best grown in early spring or for fall harvest. Sow seeds thickly every couple of weeks during cool weather and harvest with a pair of scissors when plants are about 4 inches high.

**Watercress**
***Nasturtium officinale***

This is the best-known and most widely commercially produced cress. The commercial product is grown in pure, gently running water. If you have a small stream, you can grow watercress. It is perennial where winter temperatures don't fall below -10° F.

In the home-garden adaptation of its natural growing conditions, watercress can be grown from either seed or cuttings. Make cuttings from market watercress, place them in sand or planting mix in a pot, and set the pot in a tub of water. Or sow seed in small containers and transplant the seedlings when they are 2 to 3 inches tall. Set them out in a box, a cold frame, or a trench in the soil, wherever plants can be given a continuous supply of water.

Watercress is ready for harvest in 60 days. Cut leaves after the plants bloom in spring; harvesting can continue all summer until frost.

**How to use** Cress makes a spicy, peppery, fresh addition to a tossed green salad, and an effective garnish on sandwiches or with hot or cold dishes. In Britain, cress is a very popular green often used in tea sandwiches. The French have popularized watercress soup, with a thick cream-of-potato base. Italian cooks add cress to minestrone. The Chinese have long used it in wontons. Any of the cresses can be used to flavor butter.

## CUCUMBER
## *Cucumis sativus*

The cucumber probably originated in India. Vegetable historians say that it was introduced into China during the second century B.C. The French, in 1535, found the Indians growing it in what is now Montreal, and Spanish explorer Hernando de Soto found it being grown in what is now Florida in 1539.

Because of its short growing season, 55 to 65 days from seed to picking size, the cucumber can find the warm microclimate that it needs in almost every garden. But being a warm-weather plant and very sensitive to frost, it should be direct sown only after the soil is thoroughly warmed in spring and air temperatures are 65° to 70° F.

Cucumbers respond to generous amounts of organic matter in the soil. For special treatment, dig the planting furrow 2 feet deep and fill the first foot or so with manure mixed with peat moss, compost, sawdust, or other organic material. Fill the rest of the furrow with soil, peat moss, and 5-10-10 fertilizer at 2 pounds per 50 feet of row.

To grow cucumbers in rows, leave 4 to 6 feet between the rows. Sow seed 1 inch deep, 3 to 5 seeds per foot of row. Thin the seedlings to about 12 inches apart.

To grow cucumbers in hills, space hills 2 to 3 feet apart in rows 4 to 6 feet apart. Sow 9 to 12 seeds in each hill. Thin later to 4 or 5, and finally to only 2 or 3, plants per hill. There is no specific advantage to hill planting. It probably became popular for ease of watering young plants.

Where the growing season is short, start seed indoors 4 to 6 weeks before it is time to set out transplants, which is after the frost danger has passed. Cover the transplants with hot caps or plastic to increase the temperature around the plants and protect them from unexpected late frosts.

Since cucumber roots will grow to a depth of 3 feet in normal soil, watering should be slow and deep. If the plant is under stress from lack of moisture at any time, it simply stops growing. (It will pick up again when moisture is supplied.) It is normal for leaves to wilt in the middle of the day during hot spells, but check the soil below the surface to make sure it is not dry.

When space is limited you can train cucumbers on a trellis, pole, fence, or other support. They will take very little ground space and produce

Cucumber ready for harvest

Cucumbers supported on a pole

more attractive fruits and fewer culls. Some varieties that have curved fruits when grown on the ground grow almost straight when trained on a support.

Also consider the midget, or bush, varieties when space is limited. They can be grown on the ground, in tubs and boxes, or as hanging baskets. See Cucumber Varieties on this page.

If the first early flowers fail to set fruit, don't worry. The male flowers open first; then about a week later you'll see flowers with baby cucumbers at their bases. These are the female flowers.

If this delayed setting does concern you, try one of the gynoecious, or self-pollinating, hybrids. (See Cucumber Varieties on this page.) They set with the first blossoms.

Most home gardeners have had the experience of slicing into a fresh, crisp green cucumber only to find the flesh too bitter to eat. A lack of or variation in soil moisture during the growing season has often been said to cause bitterness. Some growers feel that it is more prevalent during cool growing seasons than warm ones. Faulty fertilization, harvesting during the wrong part of the day, and peeling in the wrong direction are also thought by some to contribute to the problem. Bitterness is generally more concentrated at the stem end but never penetrates as deeply as the seed cavity. Usually it is just under the skin and can be peeled away. The direction of peeling has no bearing on either the amount of bitterness or the amount of flesh that has to be removed to eliminate it.

**Harvesting** Slicing cucumbers are harvested when they are 7 to 8 inches long, pickling cucumbers when they are 1½ to 3 inches long for sweet pickles and larger for dills. Cut the cucumbers off the vine with a knife or shears to prevent breaking the stem.

Keep all fruit picked from the vines as they reach usable size. The importance of this can't be overstressed, because even a few fruits left to mature on the plant will completely stop the set of new fruit. If you can't keep up and want the fruit to keep coming, share the harvest with your neighbors.

**Cucumber Varieties**

There are many types of cucumbers and varieties of each type. On seed packets and in catalogs are the words *slicing, pickling, bush, white-* or *black-spined, burpless,* and *gynoecious.* There's also reference to disease resistance.

Catalogs divide cucumbers into slicing and pickling varieties. Sometimes you'll see one labeled "dual purpose." Slicing cucumbers are cylindrical and grow up to 10 inches long. Pickling cucumbers are shorter and blockier. Some picklers can be picked at any age, meaning small ones for sweet pickles and larger ones for dills. It is true that all cucumbers should be picked in the immature stage, but a slicing variety that is just right at 8 inches long is not really of pickling quality at the small, sweet-pickle size.

Bush cucumbers, which do not develop long vines but instead are compact plants, are a fairly new development. Some of those available trace their parentage to a very small dwarf parent type developed many years ago by A. E. Hutchins at the University of Minnesota. Bush and semibush cucumbers types have seen more use by pickle manufacturing companies, who use mechanical harvesters, than by home gardeners. Only recently have many bush varieties been made available to the home gardener.

White-spined or black-spined has nothing to do with cucumber quality when picked. The white-spined variety merely may have more eye appeal as the spines disappear. The spines are the miniature stickers that protrude from the warts when fruits are young. White-spined cucumbers turn creamy white when old; black-spined varieties turn yellowish orange. Most gardeners don't leave cucumbers on the vine long enough to see which color the spines turn.

*Gynoecious,* a word that might slow you up, simply refers to varieties that have almost all female flowers. In a regular cucumber, which is monoecious, male blossoms greatly outnumber the female, or fruiting, blossoms. There is no delayed setting of fruit

Lemon cucumber

Pickling cucumbers

with the gynoecious hybrids, which set with the first blossoms. They also set closer to the base, or crown, of the plant, have higher yields, and mature earlier.

Gynoecious cucumbers, however, need a male pollinator plant, seeds of which are included in packets of gynoecious varieties. Male plants have green seeds and female plants have beige seeds. One male plant is needed for every five to six female plants.

There are also self-fertilizing cucumbers, which set fruit without pollination and are seedless. These varieties must be grown in an isolated area to prevent cross-pollination with other varieties. They are the type grown in greenhouses, and several varieties have become available to the home gardener.

Some gardeners are lucky enough to be able to grow cucumbers without damage from their several diseases. These are anthracnose, a problem particularly in the Southeast; downy mildew, worst in Atlantic and Gulf states; and powdery mildew, cucumber mosaic, and scab—most serious in northern states. In most areas scab and mosaic can seriously reduce harvests.

The best insurance against cucumber disease is built-in resistance. Vegetable breeders have developed many resistant strains. In the following list of varieties, disease resistance or tolerance is indicated as follows: anthracnose (A), bacterial wilt (BW), downy mildew (DM), leaf spot (LS), mosaic (M), powdery mildew (PM), and scab (S). All the varieties listed here are monoecious (have both male and female flowers) unless otherwise noted.

**Slicing cucumbers**

'Armenian', 70 days, a ribbed light green novelty that can grow to 3 feet long.

'Burpee Hybrid II', 55 days, with straight 8-inch fruit, gynoecious (M, DM).

'Burpless', more digestible, may be pickled, 10- to 12-inch hybrid fruit, straight if grown on a support, 62 days, some disease tolerance. Hybrid.

'Early Surecrop', vigorous, widely adapted, 8- to 9½-inch medium-green fruit, All-America Selection, 58 days (DM, M).

'Gemini 7', developed at Clemson University in South Carolina, gynoecious, 8- to 8½-inch dark green fruit, 60 days (A, DM, LS, M, PM, S).

'Marketmore 76', developed at Cornell University in northern New York State, one of the finest main crop slicers for northern areas, 8-inch fruit, 67 days (M, S, DM, PM).

'Poinsett 76', developed at Clemson University, widely adapted, mid to early 7½-inch fruits, 63 days, highly recommended for hot climates (A, S, DM, LS, PM).

'Salad Bar', 57 days, hybrid, 8-inch fruit. Resistant to several diseases.

'Slicemaster', gynoecious hybrid, 8-inch fruit, 55 days (DM, PM, LS, A, S, M).

'Straight Eight', 63 days, slightly striped, straight 8-inch fruit. An All-America Selection.

'Super Slice', hybrid with straight, slender 9-inch fruit. 64 days (S, M).

'Sweet Slice', 10- to 12-inch very mild fruit, straight if trellis grown, All-America Selection hybrid, 63 days, burpless (DM, PM, LS, M, A, S).

'Sweet Success', self-fertilizing and seedless, burpless, 14-inch hybrid, All-America Selection winner (M, S, LS, DM).

'Victory Hybrid', gynoecious, slim, 8-inch deep green, fruit, All-America Selection winner, 60 days (A, DM, M, PM, S).

'Whopper', 55 days, crisp dark green fruit, gynoecious (S, M, LS, A, PM, DM).

**Pickling cucumbers**

'County Fair 87', 48 days, self-fertilizing, seedless hybrid with crisp 3-inch fruit that can be sliced (BW, PM, DM, S, M, A, LS).

'Earlipik', 53 days, gynoecious, hybrid, 5-inch fruit (PM, M, S).

'Liberty', North Carolina State University hybrid, excellent home garden pickler, 3 inches long, vigorous, good cold tolerance, gynoecious, All-America Selection winner, 56 days (DM, M, PM, S), also resistant to angular leaf spot and target spot.

'Lucky Strike', 52 days, compact plant, good in cooler areas, gynoecious, tolerant of most diseases. Hybrid.

Eggplant and tomatoes

Eggplant cubed for cooking

'Saladin', European origin, gynoecious, semismooth, tender, nonbitter skins, 5-inch bright green fruit, can be used for slicing, 55 days, hybrid (BW, M, PM).

**Bush cucumbers** 'Bush Champion', vines short and compact, 9- to 11-inch fruit over a long season, good for slicing, 60 days (M).

'Bush Crop', 60 days, straight, slicing hybrid variety with 7-inch fruit.

'Bush Pickle', 48 days, 4- to 5-inch fruit for pickling.

'Patio Pik', gynoecious hybrid, small vine, good yielder, 4-inch medium-green fruit best used for pickling (A, DM).

'Pot Luck', early dwarf hybrid vine for limited space, 6½- to 7-inch dark green fruit, 58 days. Gynoecious. Slicing type (M, S).

'Salad Bush', 58 days, an All-American Selection with slicing cucumbers 8 inches long (DM, PM, M, S, LS).

'Spacemaster', widely adapted dwarf plant, 7½-inch dark green fruit, 60 days (M).

**How to use** For freshness and crispness it's important not to peel or slice a cucumber until just ready to use. (Many of the new varieties needn't be peeled at all.) If you want fancy, scalloped slices, score with a fork before cutting. Soaking slices in ice water will make them even crisper for a relish tray.

This popular salad and pickling vegetable is also good served with sour cream and a sprinkling of parsley or, for a Middle-Eastern touch, with yogurt and mint. Dress slices with a vinegar-sugar-dill sauce for a Scandinavian accent. Easy-to-fix salads can be made with a mayonnaise or vinaigrette dressing. Cucumbers also make delicious cold soups perfect for hot summer days.

Although cucumbers are most often used raw, many people find them more digestible cooked. Boil slices quickly and serve with melted butter or a light cream sauce; sauté with chopped tomatoes or onions; bread and fry until golden; or bake stuffed with meat, cheese, chopped mushrooms, and bread crumbs.

## DANDELION
## *Taraxacum officinale*

This is basically the same plant that grows as a weed in lawns, but the cultivated types will grow much larger, with thicker leaves and better flavor.

The dandelion is a cool-weather vegetable, so plant in early spring or late summer for fall harvest. If you've ever collected wild dandelions, you know that the best are found in moist, fertile soil in the cool of spring.

The dandelion is easily grown from seed. It is perennial and if allowed will come up each year.

Pick greens when the leaves are still young and tender and before the plant flowers. Keeping flowers picked as soon as they bloom will increase the harvest of the leaves; the flowers can be used for wine. With advanced maturity, leaves become fibrous and tough. To slow this, tops can be blanched by tying up the leaves or by covering the plant to exclude any light.

'Thick-Leaved' is the most common variety of dandelion, with large, thick dark green leaves that are ready to harvest in 95 days.

**How to use** Called spring-tonic greens, dandelions contain more iron and vitamin A than any other garden vegetable or fruit.

The leaves, especially of the cultivated varieties, have a tangy flavor that goes well in salads with thin slices of sweet onion and tomato and chopped basil. Or the leaves can be cooked and eaten like spinach. Add them to a thick lentil soup along with bits of bacon and chopped onion. The roots are roasted and ground and used like coffee.

Dandelion wine, made from the flowers, is a time-honored tradition with many gardeners. Said to be among the better of the homemade wines, dandelion wine has a bouquet similar to that of champagne. The overall effect is memorable.

## EGGPLANT
## *Solanum melongena*

One of the oldest references to eggplant is in a fifth century Chinese book. A black dye was

Miniature eggplant

White eggplant

made from the plant and used by ladies of fashion to stain their teeth, which, after polishing, gleamed like metal.

Wild eggplant occurs in India and was first cultivated there. The Arabs took it to Spain, the Spaniards brought it to the Americas, and the Persians to Africa.

The eggplants received in various European countries in the sixteenth and seventeenth centuries varied greatly in shape and color. The first known eggplants appear to have been of the class now grown as ornamentals, the fruit white and resembling an egg. By 1806, both the purple and white ornamentals were growing in American gardens. Modern eggplants are oval, round, or elongated and have shiny purplish black fruit.

Eggplants are more susceptible to low-temperature injury, especially on cold nights, than tomatoes. Don't set plants out until daily temperatures are in the 70° F range. Plants that fail to grow because of cool weather become hardened and stunted; once stunted they seldom make the rapid growth necessary for quality fruit. If frost dates are unpredictable in your area and very late frosts common, use hot caps or plastic covers.

Because eggplant can take 150 days to mature from seed, most gardeners grow transplants started indoors 6 to 9 weeks before the date of the average last frost.

Transplant seedlings about 18 inches apart in rows 36 inches apart. The soil should be fertile and well drained. Apply a side-dressing of fertilizer in a month, and again in another month. Plants heavy with fruit may require some support. Watch out for flea beetles and the Colorado potato beetle.

Harvest the fruit 75 to 95 days after setting out the transplants. For best eating quality, select fruit that has a high gloss and pick fruit when it is young, at about one third to two thirds its normal mature size of 4 to 5 inches across. Early harvest encourages repeated flowering and fruit set. One test for maturity is to push on the side of the fruit with the ball of the thumb; if the indentation does not spring back, the fruit is mature. If upon opening the fruit the seeds are brown, the best eating stage has passed. Since the stem is woody, cut it with pruning shears, leaving some of the stem on the fruit.

Eggplant grows well in containers. Using a planting mix is good insurance against the diseases that plague eggplant in some areas. Varieties with medium- to small-sized fruits carried high on the plant are more attractive for container growing than are the low-growing, heavy-fruited types. Where summers are cool, place containers in the hot spots around the house—in the reflected heat from a south wall, for example.

**Varieties** Whereas large-fruited eggplants are the most popular market varieties, the home garden trend is toward the smaller European types. There are also Asian eggplants that are more slender and have dull skin. The days until maturity in the following varieties applies to transplants.

'Beauty Hybrid', 69 days, hybrid with round fruit and glossy black skin. Resistant to fusarium wilt and tobacco mosaic virus.

'Black Beauty', very uniform, standard eggplant, about 4 large, round, purple fruits per 18-inch plant, slightly different strains available, 70 to 80 days.

'Black Bell', oval fruit, very dark and glossy, harvestable at about 6 inches, productive 28-inch plants, 60 days.

'Dusky', one of the earliest and most productive hybrids, thin and cylindrical jet black fruit, harvestable at about 5 inches, 56 days, resistant to tobacco mosaic virus.

'Easter Egg', 52 days, ornamental, egg-shaped white fruit. Resistant to tobacco mosaic virus.

'Ichiban', Asian-type hybrid, narrow 12-inch fruit, productive plants 36 inches high, 60 days.

'Satin Beauty', 65 days, hybrid improvement of 'Black Beauty' with dark purple fruit on a compact plant.

'Tycoon', an Asian type, 54 days, a hybrid with slender purplish black fruit.

**How to use** This amazingly versatile vegetable can be steamed, baked, fried, boiled, sautéed, breaded, stuffed, or sauced. Combining well with

'Green Curled Ruffec' endive

Florence fennel

cheese, tomatoes, onions, garlic, herbs, and meats, it is the delight of cooks in many countries.

Because of its spongy texture, eggplant has a tendency to become watery. You can eliminate excess moisture by slicing, salting, and draining before use. Or stack the slices, cover with a heavily weighted plate, and let stand until the moisture is squeezed out. As delicious as eggplant is fried, it also tends to soak up oil, so if you're counting calories, it's better to bake them than fry or sauté them.

The Greek national dish, moussaka, combines chopped meat and layers of eggplant topped with a béchamel sauce and sprinkled with cheese. In India cooked chunks of eggplant are served in a curry sauce. French cooks use it in ratatouille, a blend of garden vegetables baked and served hot as a side dish or cold as a luncheon salad. Try Italian eggplant Parmesan—slices of eggplant coated in egg and bread crumbs, then baked with a topping of tomato sauce and grated cheese.

## EGYPTIAN ONIONS

See Onion Family.

## ENDIVE AND ESCAROLE *Cichorium endiva*

Endive and escarole are basically the same plant, but varieties called endive have curled and cut, lacy leaves, whereas those called escarole have smoother broad leaves. True endive should not be confused with French or Belgian endive, which is a variety of chicory (see page 84). Endive and escarole are closely related to chicory and radicchio (see page 116).

Endive and escarole are grown in the same way as lettuce and have a better, less bitter flavor as a fall crop. To reduce the bitterness and increase the tenderness of the leaves for use in salads, blanch by drawing the outer leaves together and tying them with string for 2 to 3 weeks. Or cover the plants with a bushel basket.

**Varieties** 'Broad-Leaved Batavian', 85 days, an escarole with slightly crumpled deep green leaves; 'Florida Deep Hearted', 90 days, an escarole with coarse leaves; 'Green Curled Ruffec', 95 days, an endive with curled, deeply cut medium-green leaves; 'Salad King', 100 days, an endive with curled, deeply cut leaves.

**How to use** Both endive and escarole have a slightly bitter flavor and are used as a salad green, either alone or mixed with lettuce and other greens. Both may also be cooked, but it's best to cook the outer leaves and use the inner ones for salads.

## FLORENCE FENNEL OR ANISE *Foeniculum vulgare* var. *azoricum*

The vegetable that supermarkets commonly call anise or sweet anise is known to gardeners as Florence fennel. It's also known as *fenouil* to the French and *finocchio* to the Italians; Americans call it sweet fennel or just plain fennel. Fennel was very popular among the Romans, who it is said served virtually no meats or vinegar sauces without it.

The plant grows about 2 feet high with broad, ribbed leafstalks that overlap each other at the base, forming a bulbous enlargement that is firm, sweet, and white inside. Its leaves are finely divided and look like dill but have the flavor of anise. There is a common fennel that grows 4 to 6 feet high and is often included in herb gardens. Its seeds are useful for flavoring, but it lacks the broad stalks at the base.

Florence fennel requires cool weather. Plant it either as early in spring as the soil can be worked, or in summer for a fall crop. In mild-climate areas plant in fall for a spring harvest.

Sow seeds about ¼ inch deep in a sunny spot. Thin seedlings to 18 inches apart. When the plants are half-grown, hill up the soil around the bases to blanch the leaves. Seeds started indoors should be in peat pots, since they do not like to be transplanted. The soil should be rich and alkaline.

Florence fennel will wait in the garden for a considerable time until you're ready to harvest.

Horseradish grown in sunken pot

Horseradish leaves

**How to use** Fennel has a delightful licoricelike flavor. Use it much as you would celery. Cut off the green stalks and tough outer leaves. Slice the inner leafstalks thinly, sprinkle with olive oil, and season. Chill before serving. Excellent with fish, fennel is often added to court bouillon for seafood or to a basting sauce for broiled fish. Some dieters believe that munching on raw stalks curbs their appetite.

## GARDEN HUCKLEBERRY *Solanum melanocerasum*

The garden huckleberry is not a huckleberry but an edible member of the nightshade family, therefore related to the potato, eggplant, and tomato. It grows about 2¼ feet high. The ½- to ¾-inch berries are borne in large clusters and are shiny black when ripe, in about 70 days. They are not edible until black and soft.

Huckleberry culture is the same as that for eggplant and tomato. Start seeds indoors; after all danger of frost has passed, transplant seedlings to the garden 2 feet apart in rows 3 feet apart.

**How to use** The garden huckleberry sometimes has a bitter taste, which can be removed by parboiling for 10 minutes in water containing a pinch of baking soda. When combined with lemons, apples, or grapes, huckleberries make excellent jellies, preserves, and pies.

**Recipe for huckleberry pie** Wash and stem 2½ quarts of berries. Cover with water and let come to a boil, then add ½ teaspoon baking soda and boil 1 minute; drain. Add 1 cup cold water to berries and cook until soft. Mash berries, then add 1½ cups sugar and the juice of ½ lemon, and boil about 15 minutes. Remove from heat and let cool. Add 1 tablespoon tapioca, then pour filling into pie crust. Dot with butter, add top crust, and bake.

## GARLIC

See Onion Family.

## GOURDS

See Squashes, Pumpkins, and Gourds.

## GREEN ONIONS OR SCALLIONS

See Onion Family.

## HONEYDEW MELONS

See Melons.

## HORSERADISH *Armoracia rusticana*

Horseradish grows naturally throughout much of Eastern Europe from the lands bordering the Caspian Sea through Russia and Poland to Finland. Planted in Colonial American gardens, it escaped to flourish as a wild plant.

Horseradish rarely produces seed and is generally grown from root cuttings. Ask for them at a local nursery, or check the list of catalog sources on page 139. Horseradish is a perennial hardy to -10° F, but it is usually grown as an annual because old roots are tough and not as tasty as young roots.

Set each root cutting small end down and with the large end 2 to 3 inches below the soil surface. Space plants about 12 inches apart. Roots set out in spring will be harvest sized in fall. When the leaves are about a foot high, pull back the soil above the cuttings and remove all but one or two of the crown sprouts. At the same time, with gloved hands rub off the small roots from the side of the cutting. Don't disturb the branch roots at the base. Recover the root with soil. This operation should be repeated in about a month for top-quality horseradish.

Most growth occurs during late summer and early fall, so it's best to delay harvest until October or November. Dig up the entire root; pieces of root left in the ground will grow the following spring, but they will not have the quality of young roots.

**Varieties** 'Maliner Kren' is the standard horseradish variety, forming straight white roots in 150 days.

Horseradish plant in bloom

Jerusalem artichoke

**How to use** Beloved for its tangy flavor, horseradish plays a role in cuisines throughout the world. Peel and grate the root directly into white wine vinegar or distilled vinegar. (Do not use cider vinegar, which discolors horseradish within a short time.) The vinegar may be full strength or slightly diluted, as you prefer. Bottle as soon as possible after grating, and refrigerate at all times to preserve the pungent flavor. The mixture will keep for a few weeks. Horseradish may also be dried, ground to a powder, and bottled. So prepared, it will not be as high in quality as when grated fresh, but it will keep much longer.

Mix horseradish with whipped or sour cream to accompany sauerbraten, roast beef, pork, and tongue; or add it to seafood sauces for a special tang.

## HUSK TOMATO, GROUND-CHERRY, OR STRAWBERRY TOMATO *Physalis peruviana*

The husk tomato, a bushy plant about 1½ feet tall, is grown mostly in the same way as the tomato, except that it prefers cool climates. The fruit is about the same size as the cherry tomato. The husk tomato is known by a number of other common names, including cape gooseberry, ground-cherry, strawberry tomato, gooseberry tomato, cherry tomato, poha, and winter cherry. It is closely related to the tomatillo and is a native of South America.

The fruits are deep yellow and are produced inside a paperlike husk. (The ornamental Chinese lantern is also a close relative.) When ripe, the husks turn brown and the fruits drop from the plant. Fruits left in the husks will keep for several weeks, and their flavor is improved by a light frost. They are sweet, with a spicy flavor reminiscent of cinnamon and cloves.

**How to use** The fruit is sweeter than the small-fruited tomatoes and is used in pies and jams or may be dried in sugar and used like raisins. Hawaii's poha jam is husk tomato jam. Husk tomatoes may also be eaten raw.

## JERUSALEM ARTICHOKE OR SUNCHOKE *Helianthus tuberosus*

This easily grown vegetable is native to eastern North America. It bears no relation to the artichoke, except vaguely in flavor, and has no connection to the Holy Land. In 1616, explorers of the New World discovered it being eaten by the Indians and took it back to Europe under the Italian name for sunflower, *girasole* (literally, "turning to the sun"), and the French name, *artichauts de Canada.* Today the French call them *topinambours,* and the English, Jerusalem (likely a corruption of the Italian *girasole*) artichokes.

A perennial hardy to -10° F, this plant is a species of sunflower, growing 6 to 10 feet high on a single stalk, and topped by 3-inch yellow flowers. The stalks are often used as a windbreak for tender crops, or are stripped of leaves and dried to serve as stakes for pole beans the following season. But the plant is grown primarily for its round, knobby, crunchy root, or tuber, which can be used something like a potato.

Plant Jerusalem artichokes in dry, sandy, and infertile soil in rows 3 feet apart. Plant the whole tuber or cut it into pieces with two or three eyes each. Dig a furrow 4 inches deep, drop in a tuber or piece every foot, and cover. When the first foliage appears, start to hoe, more toward the plants than away, to hill up the soil around them, as with potatoes.

Tubers should be ready for harvest in 100 to 105 days. Dig them as you need them. You can leave them in the ground for storage over the winter; the plant is a very exuberant grower, and even the smallest pieces of tubers left in the ground will produce plants the following spring.

**How to use** Unlike potatoes, Jerusalem artichokes are starch-free; their carbohydrates do not convert to sugar in the body, and they can be eaten without concern by the diet-conscious and the diabetic. Use the tubers as soon as they are dug, since they can be kept well only in soil or sand.

Jicama

'Iceberg' lettuce

Jerusalem artichokes may be prepared in all the same ways as potatoes but with an important difference: They should never be overcooked. Overcooking tends to toughen the vegetable.

They are delicious raw. Add slices to salads or use as a last-minute garnish in clear soups. If you prefer to cook them, lightly boil the tubers, with or without their skins, in salted water until tender (usually 15 to 20 minutes); add them to salads with oil-and-vinegar dressing, or quarter and sauté slowly in butter until just tender.

## JICAMA, YAM BEAN, OR MEXICAN POTATO
### *Pachyrrhizus erosus*

This is an unusual and interesting vegetable, new to American cooks but grown in its native Mexico for centuries. It is now increasingly being offered in U.S. supermarkets.

The plant, a vine, is grown principally for its large tuberous root, which is vaguely turnip shaped and usually four lobed. The skin is a brownish gray and the flesh crisp and white. The flavor resembles that of water chestnuts but is sweeter. The immature pods are edible, but the leaves, ripe pods, and seeds are toxic and narcotic and should not be eaten by humans or animals.

Jicama makes an attractive ornamental, worth a place in the flower garden. Its blooms are profuse, white to lavender in color, and resemble sweet peas; its leaves are large and heart shaped.

Jicama is a tropical plant and requires at least 9 months of warm growing season for its large roots to mature. If rich, light, friable soil and 4 months of warmth are available, the roots will be small but still quite tasty.

Soaking seeds in water for 24 hours before sowing will hasten germination. Start them indoors 8 to 10 weeks before the last spring frost. Transplants should be set into the garden as soon as weather warms. Plants may be grown like pole beans, or near a trellis for support. They can be grown on the ground but require lots of space. When growth reaches about 3 feet, pinch the tips to promote horizontal branching. Tubers form as the days begin to grow shorter, and should be harvested before the first frost. Do not allow the plants to go to seed or the tubers will be small. Flowers, which appear in late summer, should be pinched out for maximum root production.

**How to use** Jicama is delicious served as an appetizer. Peel and cut into strips or thick slices, sprinkle with lime juice and salt, and arrange on a tray. Mexicans dot it with the juices of hot chiles. Add it to fruit salads or to raw vegetable salads, or slice it thinly and prepare with pan-fried potatoes.

## KALE

See Cabbage Family.

## KOHLRABI

See Cabbage Family.

## LETTUCE
### *Lactuca sativa*

Leaf lettuce, native to the Mediterranean and the Near East, is a plant of great antiquity. More than 2,500 years ago it was cultivated in the royal gardens of the Persian kings.

As lettuce grows, so grows the gardener. With lettuce, success in the full sense of the word means not only growing a quality crop but bringing it in through many months of the year in usable quantities.

If you can plan for harvesting a salad bowl combining wedges of tomatoes, slices of bell peppers and cucumbers, and several kinds of lettuce, you have arrived as a vegetable grower.

Lettuce is a cool-season vegetable. Seeds of leaf lettuce are usually sown directly as soon as the ground can be worked in early spring. Butterhead and romaine (cos) varieties can be grown from seed or transplants; if your growing season is hot and short, start them indoors and transplant them into the garden as soon as possible,

'Bibb' lettuce

'Ruby' leaf lettuce

allowing the plants to mature before hot weather arrives. Crisphead lettuce has the longest growing season and is usually grown from transplants.

Succession crops for fall harvest are sown beginning in midsummer. Shade the seedbeds if the temperature is above 80° F. Mild-winter gardeners can grow spring, fall, and winter crops.

Lettuce does not grow well in summer heat or warm soil. High temperatures cause the plants to bolt, which makes the leaves bitter and stops the growth of the plant. Some types and varieties are more heat tolerant than others (see The Four Kinds of Lettuce on this page). In general, the most heat-resistant types are the leaf and romaine lettuces. Shading them during the heat of summer will extend their harvestability.

Lettuce occupies the soil for a relatively short time, but every day must be a growing one, with an adequate supply of nutrients and moisture. If the growth of a young plant is checked by lack of nutrients, it never fully recovers. Fertilize the soil before planting with 3 to 4 pounds of 5-10-10 fertilizer per 100 square feet.

To plant head lettuce, sow seed ¼ to ½ inch deep in rows 18 to 24 inches apart. Thin to 12 to 14 inches between plants. Thinnings can be transplanted for a somewhat later harvest. You can also find transplants at a garden center, or grow your own indoors.

For leaf lettuce, sow ¼ to ½ inch deep. Thin to 4 to 6 inches between plants in the first thinning, and later to 6 to 10 inches. The final spacing depends upon how large the particular variety grows.

**Harvesting** The outer leaves of leaf lettuce can be picked at any time as they mature. Butterhead or romaine types can be harvested either by removing the outer leaves or by digging up the entire head. Crisphead varieties are picked when the center is firm.

Never let plants suffer from lack of moisture. The most critical period of water need is when the heads begin to develop.

Thinning is extremely important. If you leave two plants of head lettuce where only one should grow, you'll probably harvest two poor heads, or none. Thinning may be troublesome, but remember, you can use the thinnings.

The best part of lettuce is the light green leaves in the center of a nearly mature plant. If the row is crowded, all you get is a bunch of little, bitter outside leaves.

### The Four Kinds of Lettuce

Lettuce varieties fall loosely into four categories: crisphead, butterhead, leaf, and romaine. Each group has individual growth characteristics as well as taste characteristics. Which group you select from depends on the season in which you will be harvesting as well as on your taste preference.

**Crisphead, also known as iceberg** If there is only one lettuce in the supermarket produce display, this will probably be it. It has a tight, firm head of crisp leaves and is the most difficult lettuce to grow.

'Great Lakes', slow to bolt, cold tolerant, crisp, serrated leaves, possibly bitter in hot weather, resistant to tip burn, 82 to 90 days.

'Iceberg', crisp, tender heads of white leaves surrounded by crinkled light green leaves. It is slightly heat tolerant and resists tip burn, 85 days.

'Ithaca', developed at Cornell University, mild, nonbolting, tip-burn resistant in all seasons, may break down in the cold of late fall weather, 72 days.

**Butterhead** These are heading types in which the leaves are loosely folded. Outer leaves may be green or brownish, inner leaves cream or butter colored. Butterhead types are not favored commercially because they bruise and tear easily, but that's not a problem in the home garden.

'Bibb', small head 3½ inches across, small, loosely folded dark green leaves, bolts in warm weather, resistant to tip burn, 75 days.

'Buttercrunch', more vigorous than 'Bibb', thick dark green leaves, firm head, heat resistant and slow to bolt, an All-America Selection, 75 days.

'Dark Green Boston', leaves substantial, loosely folded, dark green, 73 to 80 days.

'Summer Bibb', quality of 'Bibb' but slow to bolt, 77 days.

'Tom Thumb', a miniature with tennis ball–sized heads, 65 days.

**Leaf or bunching** These are open in growth and do not

Row of butterhead lettuce

Assorted melons

form heads. Many variations are common in the outer leaves; some are frilled and crumpled, some deeply lobed. Leaf color varies from light green to red and brownish red. Leaf lettuce matures quickly and is the easiest lettuce of all to grow.

'Black-Seeded Simpson', moderately crinkled light green leaves with curled margins, fast growing and heat tolerant, 44 days.

'Grand Rapids', frilled and crinkled light green leaves, resistant to tip burn, slow to bolt, 45 days.

'Green Ice', juicy, extracrisp, wavy-margined, glossy dark green leaves. One of the slowest to bolt, 45 days.

'Oakleaf', medium-sized plants with thin, deeply lobed leaves with a sweet flavor, heat resistant, 40 days.

'Prizehead' (or 'Bronze Leaf'), vigorous, mild in flavor, large, broad, slightly frilled bronze-tinted leaves, heat tolerant, 45 days.

'Red Sails', ruffled and fringed reddish bronze leaves, the reddest of any leaf lettuce, slow to bolt, an All-America Selection, 42 days.

'Red Salad Bowl', as attractive as it is delicious, deeply lobed bronze red leaves, crisp, easily grown, heat resistant, 50 days.

'Ruby', deserves high marks for color, crinkled light green leaves with bright red markings, heat resistant, an All-America Selection, 50 days.

'Salad Bowl', tender, crinkly, deeply lobed lime green leaves in a broad clump, heat resistant and slow to bolt, an All-America Selection, 40 days.

'Slobolt', a 'Grand Rapids' type that takes more heat than 'Grand Rapids', 45 days.

**Romaine or cos** These lettuces grow more upright and cylindrical, to 8 or 9 inches high. The leaves are tightly folded, medium green on the outside and greenish white inside. They taste sweeter than other types of lettuce.

'Dark Green Cos', upright, thick, slightly crinkled dark green leaves, 70 days.

'Paris Island Cos', with slightly crinkled medium green leaves in a tight head, slow to bolt and resistant to tip burn, 70 days.

'Valmaine' is similar to 'Paris Island Cos' but is tolerant of downy mildew.

**How to use** It may seem almost unthinkable to use lettuce any other way than rinsed, patted dry, chilled, and served with your favorite salad dressing, either solo or with other vegetables, or used as a bed for chicken and tuna salad or as a finishing touch to sandwiches. However, if your garden blesses you with an overabundance, try something different: Braise it in butter and flavor with nutmeg, as French cooks do; or make cream of lettuce soup, flavored with a dash of curry and garnished with chopped, hard-cooked egg. You can also try lettuce in wilted salads, stir-fried with mushrooms and onions, or as a cooked vegetable, steamed in chicken broth.

## MELONS

Melon refers to cantaloupe, muskmelon, winter melon, and watermelon. The true cantaloupe is not grown in North America, but the term is used generally to describe the early shipping types of muskmelons. Although the name muskmelon commonly refers to all types except watermelon, it may mean more specifically the melons with a musky flavor that have orange flesh and netting on the rind.

The winter melons—casabas, Crenshaws, Persians, and honeydews—are late-maturing varieties of muskmelon.

The watermelon is a different species and requires more summer heat than muskmelon.

To grow melons, first work 5-10-10 fertilizer into the soil at the rate of 4 pounds per 100 feet of row, adding generous amounts of organic matter if the soil is heavy and drains poorly. When the runners are 12 to 18 inches long, fertilize again, spreading it 8 inches away from the plants. Make a third fertilizer application after the first melons are set on the vine.

Melons do need space; plants can grow 10 feet across. Give them 12 inches between plants in rows 4 to 6 feet apart. (Some bush varieties can be spaced much more closely.)

Vines require plenty of moisture when growing vigorously, and until the melons are fully sized, but hold back on watering during the

Cantaloupe

Watermelon

ripening period to retain the melon's full flavor.

**Prescription for Success**

- Select disease-resistant melon varieties.
- In short-season or cool-summer areas, start seeds indoors 3 to 4 weeks before the outdoor planting date, or buy transplants.
- If you live in a short-season area, look for extra-early varieties.
- Use a black plastic mulch to raise the soil temperature, conserve soil water, and prevent weeds.
- Protect transplants with hot caps or row covers.

**Cantaloupes or Muskmelons**
***Cucumis melo, Reticulatus*** **Group**

The oldest supposed record of muskmelons dates back to 2400 B.C. in Egypt. The first muskmelons introduced into Europe are said to have come from Egypt to Rome in the first century A.D. Columbus reported on his second voyage to the New World that he found them growing in the Galapagos from a planting two months earlier; the planting date has been established as March 29, 1494. Muskmelons were recorded in what is now Mississippi and Alabama in 1582 and in Virginia and along the Hudson River in New York in 1609.

Muskmelons require from 70 to 100 days from seed to maturity. Gardeners in short-season or cool-summer areas should start transplants in peat pots or other containers 3 to 4 weeks before it is time to plant out, or purchase transplants at the garden center. Where the growing season is long enough, seeds may be sown into the garden 2 weeks after the last spring frost.

Some gardeners may manage to pick muskmelon at just the right time, but it is difficult to know exactly when a melon is ripe. A general guide is that melons that will be shipped to market are usually harvested at the full-slip stage, meaning that the stem breaks away cleanly with slight pressure. Vine ripe is when the stem breaks cleanly when you just lift the melon.

**Varieties** Disease resistance in these recommended varieties is indicated as follows: fusarium wilt (F), powdery mildew (PM).

'Ambrosia', extrasweet, juicy salmon flesh, small seed cavity, 86 days (PM).

'Burpee's Hybrid', thick, sweet deep orange flesh, vigorous, 82 days.

'Earlisweet', deep salmon flesh and thick, numerous 5-inch fruits, good for short growing seasons, 70 days.

'Hale's Best Jumbo', close and heavy netting, gray-gold rind at maturity, slightly oval 6½- to 7½-inch fruit, firm flesh, 87 days.

'Honeybush', bush-type plants, bright salmon flesh, 82 days (F).

'Minnesota Midget', 4-inch fruit, very compact vines requiring only 3 feet of space, 60 days.

'Samson Hybrid', heavy fruit with a high sugar content and deep orange flesh, All-America Selection, 85 days (F, PM).

'Super Market', deep salmon flesh, sandstone rind at maturity, 6 to 7 inches, 84 days (F, PM).

'Sweet 'n Early', ripens early, bears for an extended period, flavorful salmon flesh, 75 days (PM).

**Watermelon**
***Citrullus lanatus***

The culture of the watermelon in North Africa is believed to go back to prehistory. Watermelon is of great antiquity in the Mediterranean lands and has been cultivated in Russia, the Near East, and the Middle East for thousands of years.

The European colonists brought watermelon seeds to North America, and the plant is recorded as "abounding" in 1629 in what is now Massachusetts.

Watermelon requires more summer heat than muskmelon. They especially require warm nights. In areas that are not hot enough to grow 25- to 30-pound watermelons, but where muskmelons can be grown, the medium-sized icebox-type watermelons are the best bet. Most watermelons have red flesh, but varieties also exist with yellow and white flesh.

Watermelon is grown in the same way as other melons. Final spacing of standard vining varieties should be 3 feet apart in rows 4 feet apart.

Picking a watermelon when it's neither too green nor too ripe is not easy. Some claim ripeness when the little

Casaba

Honeydew on aluminum reflector for added heat

pig's-tail curl at the point of attachment to the vine turns brown and dries up; but in some varieties it dries up 7 to 10 days before the fruit is fully ripe. The sound of a thump—a ringing sound if the fruit is green or a dull or dead one when the fruit is ripe—is also unreliable because a dull, dead sound is also the sign of overripeness.

The surest sign of ripeness in most varieties is the color of the bottom surface. As the melon matures, the ground spot turns from the color of light straw to a richer yellow. In addition, most watermelon tends to lose the powdery or slick appearance of the top surface, becoming more dull when ripe.

**Varieties** 'Bush Baby II Hybrid', round 10-pound fruits, bright red flesh, dwarf plants, 80 days (F and anthracnose).

'Charleston Gray', red flesh, grayish green rind, oblong 28- to 35-pound melons, 85 days (F and anthracnose).

'Crimson Sweet', for full-season areas, 15- to 25-pound fruits, rind striped with dark green, dark red flesh, averaging 11 percent sugar, 80 days (F and anthracnose).

'Sugar Baby', early and productive icebox-type melon, green stripes turning almost black when ripe, 7- to 8-inch round fruits, medium red flesh, 8 pounds, 75 days.

'Sweet Favorite', All-America Selection, oblong 20-pound melons, bright scarlet flesh, matures in 80 days.

'Yellow Baby', very vigorous even in short-season areas, rind thin but hard, light green with dark stripes, bright yellow flesh, 7-inch oval-to-round fruits, 10 pounds, an All-America Selection, 86 days.

**How to use** Nothing is more refreshing on a summer day than a slice of chilled, ripe melon. A ripe melon will keep in the refrigerator about one week; but always wrap a cut melon to prevent its odor from permeating other foods.

Melons alone are delicious, but they can be dressed up as well. Serve with a wedge of lemon or lime; a squeeze of juice enhances the flavor. Fill muskmelon halves with fruit, ice cream, or yogurt; or freeze melon balls or cubes in light syrup and serve in a fruit compote. Peel a cantaloupe or honeydew, cut off one end, remove the seeds, then fill with a fruit gelatin mixture and allow to set. Slice and serve on lettuce leaves.

After a watermelon is enjoyed, don't discard the rind. Use it to make pickled watermelon rind.

**Winter Melons**
***Cucumis melo,***
***Inodorus* Group**

These would be more aptly named long, hot summer melons. The casaba, Crenshaw, Persian, and honeydew melons find their best growing conditions in the hot interior valleys of California, Arizona, and the Southwest. They are called winter melons because they mature late.

The casaba is wrinkled and golden when mature, with very sweet and juicy white flesh. A good keeper, the casaba appears in markets for months after harvest season. Try 'Golden Beauty', white flesh with a golden rind at maturity, 110 days.

The Crenshaw has salmon-pink flesh and a dark green rind that turns yellow at maturity. It is famous for its distinctive flavor. A tender rind makes shipping a problem, but that's not a concern to the home gardener. Crenshaws mature in about 110 days. 'Burpee Early Hybrid' matures in 90 days and can be grown in shorter growing seasons than most Crenshaws.

The Persian melon is larger than most muskmelons, round, and heavily netted with thick orange flesh. It needs about 95 days in a hot, dry climate.

The rind of the honeydew is creamy white, smooth, and hard. The flesh is lime green, with a slight golden tinge at maturity, which arrives at about 112 days. 'Earli-Dew' is a hybrid with lime green flesh and a creamy green rind, 80 days, good for short growing seasons.

Winter melons are grown and harvested in the same way as muskmelons but need a longer, hotter growing season. Using clear plastic mulch, which traps more heat than black plastic, will help when growing them in shorter,

Mustard greens

Nasturtium

cooler seasons than they demand. So will growing them near a heat-reflecting wall.

## MUSTARD GREENS
## *Brassica juncea* and *B. campestris*

Of the many mustards, the species most frequently grown commercially for greens is *Brassica juncea,* native to the Orient. Several varieties differing in leaf shape and texture—either cut, crisped, frilled, or curled—are offered in seed packets. Mustard greens have a peppery flavor. Another species, *B. rapa,* is listed in seed catalogs as mustard spinach or tendergreen mustard. The mustard from which the popular condiment is made is *B. nigra.*

Mustard is a cool-weather, short-day crop and bolts very early in the spring. Plant as soon as the soil can be worked in spring; in mild-weather areas it can also be planted in the fall.

Treat the same as lettuce. Sow seed in rows 12 to 18 inches apart and thin seedlings to 4 to 8 inches apart. For tender leaves, give the plants plenty of fertilizer and water and harvest before full grown. Mustard grows fast in fertile soil, 25 to 40 days from seed to harvest.

Harvest the outside leaves when they are 3 to 4 inches long and still tender, leaving the inner leaves to develop, or harvest the entire plant when warm weather sets in.

**Varieties** 'Florida Broadleaf', large, thick green leaves with a whitish midrib, easy to clean, 50 days.

'Fordhook Fancy', the dark green leaves are deeply curled, fringed, and curved backward, slow to bolt, with a mild flavor, 40 days.

'Southern Giant Curled', upright growth, leaves large and wide, curly edges, bright green tinged with yellow, slow to bolt, mild flavor, All-America Selection, 50 days.

'Tendergreen', or 'Mustard Spinach', produces a large rosette of thick, smooth, glossy dark green leaves, one of the mildest in flavor, good heat resistance, 35 days.

**How to use** Strong and distinctive in flavor, mustard greens are excellent served alone or mixed with milder greens, such as chard. Cook them quickly in a little boiling water and dress with olive oil and white wine vinegar. For a tangy salad combine them with beet greens, minced onion, hard-cooked egg, and mayonnaise. Try cooking them with a mixture of sautéed bacon and onion; or simmer salt pork or ham hocks for several hours, add greens, and let cook another half hour.

## NASTURTIUM
## *Tropaeolum majus*

If you have grown nasturtiums only for their contribution to the beauty of the garden, you are missing the important bonus they offer the good cook.

Nasturtiums are very easy to grow, bloom profusely in ordinary, well-drained soil, and even thrive in dry, sandy, or gravelly areas. Too-rich soil may cause plants to produce more leaves than flowers.

Nasturtiums do not transplant well and should be direct seeded into the garden after all danger of frost has passed. They grow best where nights are cool.

Snip off the stems at the base whenever fresh foliage, flowers, or buds are needed.

**Varieties** Which nasturtium variety should you choose? Any one will meet kitchen requirements. There are climbing types that reach 3 to 8 feet in height; semivining types that grow to 2 feet, and dwarf varieties that reach 12 inches. If you intend to plant nasturtiums in the vegetable garden, training the long vines on chicken wire will take the least space. Semivining nasturtiums, such as 'Glorious Gleam Mixture,' can be grown in hanging baskets. For boxes and pots use one of the low-growing dwarf varieties. The foot-high 'Jewel Mixed' offers a choice of color from semi-double flowers. And if you like the simplicity of the old-fashioned kinds, there's a dwarf single nasturtium available.

**How to use** The lively, subtle, peppery taste of nasturtium leaves is reminiscent of watercress. Chop and add them to all kinds of salads. Flowers can serve as carriers of cheese mixtures or as a

Okra

Bulbing onions, green onions, shallots, leeks, and garlic

salad garnish for an hors d'oeuvres tray. The plump, green, unripened seedpods are often pickled in vinegar and substituted for capers.

## OKRA
## *Abelmoschus esculentus*

Okra, sometimes called gumbo (though the name gumbo is more properly applied to soups containing okra), achieved its popularity in the French cookery of Louisiana and was probably introduced to that region by French colonists in the early 1700s.

There are two types of okra: the tall growers, from 4 to 7 feet; and the so-called dwarfs, to about 3 feet. An attractive plant, okra is an edible form of hibiscus that has yellow hollyhocklike flowers. A half-dozen plants will provide more than enough fruit over a long season for the cupful you'll need now and then for various dishes, and also leave enough for pickling.

Okra is a warm-season vegetable. Planting dates and fertilizing and watering schedules are the same as for corn. Soak seeds for 24 hours before sowing to speed germination. Plant seeds 8 to 10 inches apart in rows 3 to 4 feet apart. In short-season areas start seeds in small pots about 5 weeks before you would plant corn or beans, and set out seedlings when the soil is thoroughly warm.

**Harvesting** To keep the plant producing, no pods should be allowed to ripen on the stalk. Young pods, from 3 to 5 inches long, are more tender and more nutritious than older pods. Since the pods develop rapidly, the plants should be picked over at least every second day. Handle pods with care; broken or bruised pods are apt to become slimy or pasty during cooking.

**Varieties** 'Annie Oakley', a compact hybrid with long, slender pods that grows well in cooler areas, 52 days; 'Blondy', an All-America Selection, early, dwarf okra good for short seasons, 55 days; 'Burgundy', an All-America Selection, grows 4 feet high and has early maturing red pods that retain excellent eating quality even when mature, 60 days; 'Clemson Spineless', grows to 4 to 5 feet and is a heavy yielder of slightly grooved pods, an All-America Selection, 56 days.

**How to use** Whether steamed, boiled, baked, or fried, okra should be cooked rapidly to preserve its flavor and prevent its developing a slimy consistency. Do not cook it in iron, copper, or brass utensils. Although the chemical reaction is harmless, it will discolor the pods.

The young, tender pods of okra are very popular in Creole cooking and are excellent in soups and stews. They possess a natural thickening agent, evidenced in the preparation of gumbos. Okra combines well with other vegetables, especially tomatoes. It can also be pickled.

Pods picked at maturity can be dried and used in floral arrangements.

## ONION FAMILY

The onion and its pungent relatives have been highly regarded since antiquity. Onions fed the sweating builders of the pyramids and the conquering troops of Alexander the Great. General Ulysses S. Grant, in a dispatch to the U.S. War Department, wrote, "I will not move my armies without onions." Emperor Nero earned the nickname leek-throated because of his frequent munching on leeks to clear his throat.

An enthusiastic nineteenth century gourmet said it all for onion lovers everywhere: "Without onions there would be no gastronomic art. Banish it from the kitchen and all pleasure of eating flies with it . . . its absence reduces the rarest dainty to insipidity, and the diner to despair."

The onion family contains the familiar bulbing onions, chives, Egyptian onions, garlic, leeks, green onions, and shallots.

### *Allium cepa, Cepa* Group

The nature of the onion is to grow tops in cool weather and form bulbs in warm weather, but the timing of the bulbing is controlled by both temperature and day length. Onions are so sensitive to day length that they are divided into short-day and long-day varieties. It is very important to use varieties designated for your area.

Onion field

Setting out garlic transplants

Short-day types are planted in the southern parts of the United States as a winter crop, started in fall. They make bulbs as days lengthen to about 12 hours in early summer. When grown in the north in summer, they make only small bulbs, which can be used for pickling or cocktail onions. Long-day onions are grown in the northern latitudes. Most require 14 to 16 hours of daylight to form bulbs. Planted in spring as early as the soil can be worked, they bulb when days are longest in summer. Long-day onions planted in the South can be used for green onions but do not form bulbs.

Onions are heavy feeders, so work manure and fertilizer into the soil before planting. A pound of manure per square foot and 4 to 5 pounds of 5-10-10 fertilizer per 100 square feet will do the job. A steady moisture supply is essential, particularly during bulb formation.

Start onions from seeds, transplants, or sets (small dry onions available in late winter and early spring). Seed is generally the least popular, except for starting transplants. For transplants, sow seed indoors 12 weeks before the outdoor planting date. Sow ½ inch deep.

In the garden, space rows 1 to 2 feet apart, and thin the seedlings 2 to 3 inches apart.

Transplants are popular because of the large bulb size produced over a short time (65 days or less). Transplants are available from local garden centers.

Sets are usually the most reliable, although the varieties available are more limited. Select sets early, when they are firm and dormant. Sets are available in three colors—white, red, and brown. Most gardeners prefer white sets for green onions, but the other two colors are acceptable. Sets are generally available only for long-day onions. Divide the sets into two groups, those smaller than a dime in diameter and those larger than a dime. Use large-sized sets for green onions; the large sets may bolt and not produce good dry bulbs. Plant the small-sized sets for dry onion production, since they will not likely bolt.

Harvest by simply pulling the onions from the ground when half of the tops have broken over naturally. When the tops have fully wilted, cut them off 1½ inches above the bulb. Prepare for storage by drying in an open crate or mesh bag for 2 weeks or more. Clean by removing dirt and loose, dry outer skins. Store where the air is dry and between 35° and 50° F.

**Varieties** These recommended varieties include the important factor of climate adaptation as follows: long day (LD), short day (SD). The best onions for storage are usually those with the most pungent flavor.

'Crystal White Wax', white flesh, soft and mild, small, round onions for pickling, short storage, day length neutral, 60 days.

'Early Yellow Globe', medium sized, firm white flesh, pungent flavor, good keeper, 100 days (LD).

'Granex Yellow', large flat globes, yellow skin, white flesh, mild flavor, fair keeper, 170 days (SD).

'Southport Red Globe', red, medium-sized globes, good keeper, 110 days (LD).

'Southport White Globe', grown for green bunching onions after 65 days or a high-quality large white onion in fall, stores well, 108 days (LD).

'Sweet Sandwich', flat to globe-shaped yellow onion, very sweet, stores well, 110 days (LD).

'Texas Grano', large onion, mild white flesh, does not store well, 170 days (SD).

'Walla Walla', flat globes, mild flavor, best planted in the summer and harvested during the following summer, 300 days (LD).

'White Sweet Spanish', largest white onion, firm flesh, sweet and mild, medium keeper, 120 days (LD).

'Yellow Bermuda', large flat bulbs, soft and mild, short keeper, 185 days (SD).

'Yellow Sweet Spanish', large yellow globes, sweet and mild, medium keeper, good slicer, thrips resistant, 120 days (LD).

**How to use** Overcooked onions give off an unpleasant sulfurous odor, so don't cook them over high heat or too long. If you have to prepare lots of onions, drop them into

Chives in bloom

Garlic chives

boiling water for about 10 seconds, then drain and chill; the skins will slip off easily. Peeling onions under running water helps prevent tears; to remove the odor from your hands, rub them with salt or vinegar.

An onion by itself is delicious, but try small, whole ones parboiled and added to a medium cream sauce; or glaze them in honey and serve with pork. Scallop onion slices in the manner of potatoes, bake, then sprinkle with grated Parmesan cheese.

Larger onions can be stuffed with sausage, beef, chicken, fish, rice, or bread crumbs, and baked. George Washington is said to have favored mincemeat stuffing.

**Chives**
***Allium schoenoprasum***

No vegetable gardener should refuse to give chives room. This perennial, hardy to -40° F, can be clipped almost continuously, and a half-dozen plants will supply enough snippings for year-round use. If not clipped, chives produce pompons of lavender flowers in late spring above their grass-like, hollow leaves.

Chives grow best in rich, moist soil in full sun but will tolerate filtered shade. The easiest way to a quick harvest is to buy plants, but you can start seeds in small pots. Set plants into the garden 6 to 8 inches apart. Divide them every 3 to 4 years and fertilize every spring.

Look also for garlic chives, or Chinese chives (*Allium tuberosum*). They have a mild garlic flavor and grow like regular chives, but they are taller and have white flowers in late summer. Plant them 12 inches apart and divide yearly, since they grow fast. Try them in salads and stir-fried dishes.

**How to use** Chives impart a delicate onion flavor to a wide variety of dishes. Snip them into eggs, soups, sauces, cheese spreads, and dips. Sprinkle them into green salads or use to garnish cottage cheese or quiche. Spread chive butter on steaks or broiled seafood.

Chives are best used fresh but are almost as good frozen and are still good dried. They are highly perishable, so don't add them to food until just ready to serve, and don't put them in uncooked dishes that will be stored.

Chive flowers can be used as a garnish or mixed with white vinegar, which soon takes on a rosy hue and an oniony flavor.

**Egyptian Onions**
***Allium cepa,***
***Proliferum* Group**

Very hardy and unique, Egyptian onions can be planted in the fall throughout the country and in early spring used as green or bunching onions. By midsummer or fall they begin forming miniature bulbs at the tips of their stems, where most onions form flowers. Collect the bulbs when the tops begin to wilt and dry.

**How to use** These mild-flavored onions are an excellent choice for pickling. You may also freeze them or use them fresh. The hollow stalks are ideal for stuffing, as you would celery.

**Garlic**
***Allium sativum***

Two types of garlic are available: the type most commonly found at the market—a bulb containing about 10 small cloves; and 'Elephant Garlic', which is about six times larger, weighing up to a pound, and with a slightly milder flavor.

Both types are grown in the same way: In all but the coldest areas, set out cloves in the fall 1 inch deep and 2 to 4 inches apart in rows 1 to 1½ feet apart, with the pointed end up. In the coldest areas, plant in spring. You can buy cloves from garden centers, mail-order nurseries (see page 139), and produce markets.

Soil for garlic should be rich, loose, sandy, and well drained. Use a low-nitrogen fertilizer at planting time and when the tops are 6 inches high; too much nitrogen causes foliage growth but few and smaller new bulbs.

Harvest garlic when the tops turn yellow and fall over, 90 or more days after the cloves are planted. Dig the bulbs from the ground carefully, allow them to dry in the sun for a few days, then braid the tops into strings or tie in bunches and hang in a cool, dry place.

Green onions

Leeks

**How to use** Add minced garlic to salad dressings and meat sauces. Garlic in melted butter becomes a sauce for lobster, snails, mushrooms, and many green vegetables.

Insert slivers of garlic into lamb or beef roasts, or use cloves to season slow-cooking stews. Halve a clove to rub inside a salad bowl, or steep cloves in wine vinegar to flavor the vinegar.

Place peeled cloves in a jar or crock and cover with olive oil; seal and store in the refrigerator. The oil takes on a pungency appropriate to many dishes, the cloves are ready to use right from the jar and thus preserved, the cloves will keep for several months.

### Green Onions or Scallions
***Allium cepa, Cepa* Group; *A. fistulosum***

Green onions are either immature bulbing onions (*A. cepa*) or onions called bunching onions (*A. fistulosum*), a perennial type that has multiple tubular stems that grow in a bunch throughout the growing season, and has either small or no bulbs.

Any variety of the standard onion can be used as a green onion if it is harvested when the bulb is small. Bunching onions, more widely grown for green onions, include seed varieties such as 'Beltsville Bunching', mild flavored, most heat resistant, 120 days; 'Evergreen' ('Nebuka'), with longer and more slender stems than 'Beltsville Bunching', 120 days; and 'Japanese Bunching', long, slender stems, no bulbs, 110 days. All are winter hardy. Shoots are crisp and mild early in the season, more pungent later.

Green onions also can be grown from sets. Yellow sets are primarily from the variety 'Ebenezer', which can either be harvested for green onions in about a month or left to form standard onions; 'White Lisbon' forms white stalks in 60 days. The commercial green onion is always from a white variety.

**How to use** Green onions make an attractive garnish for soups, egg dishes, and poultry. Enjoy whole green onions cooked: Simmer in 1 inch of water for 3 to 4 minutes, then serve with butter and lemon juice; or chop them finely and add to your favorite quiche recipe. Try them in Asian foods—soups, sukiyaki, and stir-fried dishes.

Many cooks prefer green onions to standard onions in salads for their more delicate flavor. Green onions are delicious steamed and drained, then placed on toast and covered with melted butter or hollandaise sauce.

### Leeks
***Allium ampeloprasum, Porrum* Group**

Leeks have a sweeter, milder, more delicate flavor than onions. They do not form bulbs the way onions do but instead develop thick, cylindrical 9- to 18-inch stalks.

Leeks take a good 80 days to grow from transplants and up to 140 days from seed. When growing from seed, sow indoors in late winter and move to the garden in early spring. Set plants 3 inches apart. Leeks take a long time to mature; therefore, sowing seeds directly into the garden is rarely successful.

To get long white stems, plant leeks in trenches 4 to 6 inches deep and hill the soil against the stems after the plants are fairly well grown. The soil should be fertile and kept constantly moist. Fertilize at planting time and again when the tops are 6 inches high. Harvest in the fall anytime after the stalks are ¾ inch wide; they will be sweeter if allowed to grow to 1½ inches across. Plants can be left in the ground all winter if the temperature does not drop below 10° F.

**Varieties** 'Large American Flag' ('Broad London'), the most cold tolerant, 130 days; 'Titan', stalks longer and thicker than 'Large American Flag', 100 days.

**How to use** Leeks need to be washed thoroughly, since sandy grit burrows deep inside them. Trim the ends, slice lengthwise, and then hold under running water until clean.

In France this versatile onion is known as the asparagus of the poor. Steam it like asparagus or braise like celery, and serve au gratin or with a cream sauce. Try hot puréed leeks garnished with parsley, or serve a spicy leek pie as a main course. For a

Shallots

Shelled peanuts

delicious salad, chill cooked leeks and serve with hard-cooked egg, chopped parsley, and a vinaigrette dressing.

**Shallots**
***Allium cepa,***
***Aggregatum* Group**

French knights returning from the Crusades are credited with introducing shallots into Europe. They are heavily used in French gourmet cooking and have a delicate, subtle flavor.

The shallot is a multiplier type of onion, dividing into a clump of 3 or 4 cloves similar in appearance to garlic but smaller and brown instead of white. Most varieties set no seed. Both the bulbs and the foliage can be eaten, but shallots are grown primarily for their bulbs.

Shallots are hardy and will overwinter as perennials; for better results, lift the clusters of bulbs at the end of each growing season (about 90 days) and replant the smaller ones in the fall. Plant cloves 1 inch deep, 2 to 4 inches apart, and leave 12 to 18 inches between rows. Soil should be rich, loose, and neutral. Add a low-nitrogen fertilizer at planting time and some more in 6 weeks.

Harvest shallots when the tops turn brown and die down in late summer. Immature shallots can be used as you would green onions.

**How to use** Shallots have a distinctive flavor somewhere between that of onions and garlic and are highly prized in wine cookery. Many chefs consider bercy sauce (shallots, wine, and butter) the very best dressing for steaks and broiled meats.

It's important to mince shallots finely before sautéing so they will cook quickly. (Never let them brown.) In most recipes you can substitute 3 to 4 shallots for 1 medium onion.

## PARSNIPS

See Root Crops.

## PEANUTS
## *Arachis hypogaea*

This tropical from South America was taken by the Portuguese from Brazil to West Africa. Spanish galleons carried peanuts from South America to the Philippines, from where they spread to China, Japan, and India. They found a favorable climate in North America early in its history. Thomas Jefferson wrote of their culture in Virginia in 1781.

There are two types of peanut: Virginia, with two seeds per pod; and Spanish, with two to six seeds per pod. Most plants of the Virginia type are spreading; the Spanish are bunching. However, bunching varieties of Virginia are available and better adapted to short-season areas than the spreading types.

A long, hot growing season of 110 to 120 days is required for most peanuts. If summers are cool in your area, better forget them as a crop regardless of the season length.

The strange growth habit of peanuts inspires gardeners to experiment in growing them even when they must be in containers and given special protection. The plant resembles a yellow-flowering sweet pea bush. After the female flowers wither, a stalklike structure known as a peg grows from the base of the flowers and turns downward to penetrate the soil. When the peg pushes to a depth of 1 to 2½ inches, it turns to the horizontal position and the peanuts begin to form.

The soil should be light, sandy, and neutral to slightly alkaline. Peanuts require a generous supply of calcium in the top 3 to 4 inches of soil where the pods develop. To supply calcium, foliage is often dusted with gypsum (calcium sulfate) at the time of flowering, at the rate of 2½ pounds per 100 feet of row.

Plant seeds 1 inch deep and 12 inches apart in rows 2 to 2½ feet apart after all danger of frost has passed. Shelling the seeds is not necessary but it will hasten germination. Thin the plants to stand 2 to 3 feet apart.

When the plants are about 12 inches high, mound the soil around the bases and cover with a mulch. Make sure that the plants receive a regular supply of water up to 2 weeks before harvest. Excess water at harvest time may break dormancy and cause the mature peanuts to sprout.

**Harvesting** When the plants turn yellow at the end of the

Garden peas

Cowpea seedlings

season, lift each bush carefully with a garden fork, shake free of soil, and hang the plants, with the peanuts hanging by the pegs, in a warm, airy place for a few days. Let the plants cure for 2 to 3 weeks before stripping the peanuts from them.

**How to use** Peanuts are well known and have many uses. To roast peanuts with no chance of scorching, place them unshelled in a colander or wire basket. Preheat the oven to 500° F. Place the peanuts in the oven and turn it off. When the peanuts are cool to the touch, they're ready to eat. Peanut butter is easily made in a blender with shelled roasted peanuts, vegetable oil, and perhaps some salt.

## PEAS

Many vegetables are known as peas. We've put them all together in this section in spite of the fact that some are not related to garden peas.

### Garden Peas or English Peas *Pisum sativum*

Garden peas (green peas) are known as English peas in the South to avoid confusing them with cowpeas. They must be shelled before use.

English peas are a cool-season crop. They are grown from early spring to midsummer in cool areas, and in fall, winter, and very early spring where it is warm. Sow seeds 1 to 2 inches deep and 2 inches apart as soon as the soil can be worked in spring. The soil should be rich and moist. An inoculant for legumes increases growth and yield; fungicides prevent rotting in cool, damp soil.

The low-growing varieties that do not require staking are the easiest to grow. They can be planted in rows 18 to 24 inches apart. Climbers trained on chicken wire or a trellis need 3 feet between rows, but you can plant in double rows 6 inches apart on each side of the support.

Pick peas as soon as the pods are fully swollen and round but before the peas become hard.

**Varieties** 'Alaska' is one of the earliest and most popular, hardy, prolific, lower in sugar than many, uniform ripening, good for canning, plants 24 to 28 inches, 2½-inch pods with 6 to 8 peas, fusarium wilt resistant, 56 days.

'Green Arrow', vigorous vines, 30 inches tall, 4-inch pods with 9 to 11 peas forming at the top for easy harvest, resistant to downy mildew and fusarium wilt, 70 days.

'Little Marvel', early, very sweet, plants to 18 inches, 3-inch pods with 8 peas, 63 days.

'Maestro', very early, resistant to powdery mildew, 4½-inch pods with 9 to 12 peas, ready in 61 days.

'Wando', among the most heat resistant, also cold tolerant, 30-inch plants, 3-inch pods with 7 to 8 peas, 67 days.

**How to use** When small and tender, garden peas can be eaten raw in salads. Cooked, they complement any main course. Shell them just before using, and cook quickly in about an inch of boiling water.

Peas are good simply with butter, salt, and pepper but are also enhanced by many herbs. Try steaming them with fresh mint. Combine them in a cream sauce with pearl onions, celery, or carrots; or try them with creole, orange, lemon, or wine sauces.

### Cowpeas, Black-eyed Peas, or Southern Peas *Vigna unguiculata*

Middle Asia was the home of the cowpea before it migrated to Asia Minor and Africa. Then slave traders carried it to Jamaica, and in the warm climates of the West Indies it became an important food.

Cowpeas have a more distinctive flavor than garden peas. They also require more heat to grow—warm days and warm nights—and are damaged by the slightest frosts. Plant when the soil is warm for better germination.

Sow seed ½ to 1 inch deep, 5 to 8 seeds per foot of row, 2 to 3 feet between rows, after frost danger has passed. Thin to 3 to 4 inches apart.

Use nitrogen fertilizer sparingly. A side-dressing of 5-10-10 at 3 pounds per 100 feet of row after the plants are up is usually adequate.

Pick cowpeas in the green-shell stage, when the seeds are fully developed but not yet

Cowpeas

'Sugar Snap' pea plant in bloom

hard. Or let them ripen and store as dried peas.

**Varieties** 'California Blackeye' produces large, smooth-skinned peas with good flavor, white with black spots, for fresh or dried use, vigorous, resistant to several pea diseases including fusarium wilt and nematodes, 75 days.

'Mississippi Purple', green pods turn purple when mature, 65 days.

'Mississippi Silver', a brown crowder-type pea (seeds crowded in pods) developed by the Mississippi Agricultural Extension Service, good fresh, canned, or frozen, 65 days.

'Pinkeye Purple Hull', the most popular cowpea, young peas are white with a small pink eye, occasionally makes 2 crops per plant per season, 78 days.

**How to use** Cowpeas are an honored staple of southern cooking. Pick them in the green-shell stage, then shell and cook with bacon or pork. Fresh cowpeas can also be used in almost any recipe for snap beans.

To cook dried cowpeas, soak them overnight, then simmer 1 to 2 hours with onions and bacon or salt pork. For a tasty casserole, combine cooked peas with cooked rice and bake until thoroughly heated. Tabasco sauce is the favored seasoning for this dish.

### Sugar Peas, or Snow Peas *Pisum sativum* var. *macrocarpon*

This vegetable has small peas and edible pods and should be picked for tenderness when very young, just as the peas start to form. If you miss that stage, the pods will be too tough for eating, but you can still shell them and eat the peas, which are more starchy and not as sweet as garden peas. They are grown in the same way as garden peas.

**Varieties** 'Dwarf Gray Sugar', with vines to 2½ feet, can be grown without staking, light green pods 2½ to 3 inches long, 63 days.

'Mammoth Melting Sugar', vines to 4 feet, needs support, pick when peas become just visible in the pods, resistant to fusarium wilt, 72 days.

'Oregon Sugar Pod II', productive 2-foot vines, 4-inch pods usually 2 per cluster, resistant to virus, wilt, and powdery mildew, 68 days.

'Snowbird', small plants need no support, good for short growing seasons, 3-inch pods in double or triple clusters, 58 days.

**How to use** The French call the sugar or snow pea *mange-tout*—"eat it all." It is important in both French and Asian cooking.

To prepare, simply snip off both tips and remove the string (if there is one). Overcooking destroys both flavor and texture, so cook quickly, either by stir-frying or butter-steaming. Try snow peas in soups, sukiyakis, and stir-fried combinations. Leftover snow peas are an excellent addition to salads.

### Sugar Snap Peas *Pisum sativum* var. *macrocarpon*

Like the sugar pea, or snow pea, this all-new type of pea can be eaten whole, but it differs from sugar peas in having large peas, a thicker pod, and sweeter seeds. It can be grown to full size, then snapped and eaten like a garden bean. Tall vines grow to 6 feet or more and need strong support. The pods are slightly curved, medium green, and fleshy.

Sugar snap peas are grown in the same way as garden peas. Some varieties are more susceptible to powdery mildew than garden peas.

**Varieties** 'Snappy', vigorous 6-foot vines, 4¾-inch pods with 8 or 9 peas, resistant to powdery mildew, 63 days.

'Sugar Bon', compact plant 2 feet tall, sweet 3-inch pods, resistant to powdery mildew, 56 days.

'Sugar Daddy', crispy, stringless pods, compact plants 2½ feet high, 74 days.

'Sugar Snap', crunchy 3-inch pods, vigorous 6-foot vines, wilt resistant, All-America Selection, 70 days.

'Super Sugar Mel', high yielding, crisp, thick 4-inch pods, 3-foot vines, 68 days.

**How to use** Sugar snap peas can be used in various ways. This very sweet pea is delicious raw—add it to salads or use with dips or on a relish

'Hungarian Yellow Wax', 'Serrano Chile', 'Santa Fe Grande', 'Anaheim', Jalapeno, and 'Mercury Floral Gem' chiles

'Better Belle' bell peppers

tray. Steam lightly to preserve flavor, or stir-fry. Nearer to maturity, cut the pods and cook as for green beans. Mature, they can be shelled and the peas cooked like garden peas. They are also good for freezing but not canning. Remove the strings, when present, before cooking.

## PEPPERS, SWEET AND CHILE *Capsicum annuum*

Columbus, searching for a new route to the spice-laden West Indies, was thinking of a different pepper from those he found in the New World. The dried black pepper that is ground as a seasoning comes from the berries and seeds of *Piper nigrum* which is in no way related to the peppers of the genus *Capsicum* (bell peppers and chiles) that Columbus found growing in the Indian gardens in the Caribbean. His find was described upon his return to Spain as "pepper more pungent than that of the Caucasus."

Spice-hungry Europeans immediately adopted the new vegetable. Within 50 years peppers were found growing in England; in less than a century, on Austrian crown lands. They became so common in India that some botanists thought they were native.

Bell peppers are classed as a hot-weather vegetable, but their heat requirements are not as high as generally supposed. Fruit set occurs in a rather limited range of night temperatures. Blossoms drop when night temperatures are much below 60° F or above 75° F. Bell peppers thrive in areas with daytime temperatures around 75° F and nights of 62° F. Daytime temperatures above 90° F will cause excessive blossom drop, but fruit setting will resume with the return of cooler weather. Small-fruited varieties are more tolerant of high temperatures than the larger-fruited ones.

Chiles need higher temperatures to produce a good crop than bell peppers do. In cool areas, a black plastic mulch will keep the ground warmer and improve the yield of chiles.

Peppers seem to protect themselves from overloading the plant with fruit: When a full quota of fruit is underway, new blossoms drop. When some of the peppers are harvested, the plant will again set fruit—if the weather is right.

The easiest way to start peppers is to buy transplants from a nursery. Varieties of peppers adapted to each area are generally available. Growing transplants from seed is not difficult, however. Allow 7 to 10 weeks for germination and enough growth to make a good-sized transplant. Peppers also can be direct seeded into the garden in areas with long growing seasons. The shorter the season, the more reason to choose the early varieties.

Don't set out transplants until the weather has definitely warmed—at least a week or so after the last frost. When night temperatures fall below 55° F, small plants will just sit, turn yellow, and become stunted. If there is any chance of a late frost, use hot caps or similar protection.

Give the plants space to grow: Set them 24 inches apart in rows 2 to 3 feet apart. Some gardeners set out more plants than are needed and after a few weeks of growth pull out the weaklings.

When the first blossoms open, give the plants a light application of fertilizer. Water it in well. Too much fertilizer causes lush foliage growth at the expense of fruit. Any stress from lack of moisture at flowering time may cause blossoms to drop. Add a mulch to conserve moisture and stop any weeds.

When it's time to pick the peppers, use pruning shears or a sharp knife, and leave a short piece of stem on the pepper. Bell peppers are usually picked green and immature but full sized and firm; if they ripen on the plant they will be sweeter and higher in vitamin content. Chiles are picked at full maturity.

### Bell Peppers

The common bell peppers are perhaps the most familiar peppers in the United States. Most but not all bell peppers are bell shaped; some are tapered. There are also a few hot bell peppers, so don't presume that a bell pepper is sweet.

'Gold Crest' bell peppers

'Big Bertha' bell peppers

Most bell peppers that are sold green in the market will turn red, some yellow, and some purple, at maturity. All mature in 70 to 80 days from transplants; earlier maturing varieties are good for northern gardens and short seasons.

**Varieties** Resistance to tobacco mosaic virus is indicated by TM in the following varieties:

'Bell Boy', plants 1½ to 2 feet tall with good leaf cover, medium-long four-lobed, bell-shaped, thick-walled fruits glossy green then deep red, All-America Selection hybrid, 75 days (TM).

'Better Belle', similar in appearance and characteristics to 'Bell Boy' but maturing earlier, at 65 days.

'Big Bertha', upright 2½-foot hybrid plants are dark green with a thick canopy and have three- or four-lobed deep green to red fruits with very thick walls. This is the largest bell pepper at 7 by 4 inches, excellent for stuffing, grows well in cool climates, 72 days (TM).

'California Wonder', good for stuffing, four-lobed bell pepper with tender flesh and delicate flavor, fruits dark green then red, 4 by 4 inches, 73 days.

'Canape', hybrid good for short growing seasons and resistant to heat and drought as well. Fruits are three-lobed, tapered, thick walled, and 2½ by 3½ inches, 62 days (TM).

'Cherry Sweet' (also called 'Red Cherry Sweet'), a small, rounded to slightly tapered 1½-inch pepper good for pickling, 78 days.

'Cubanelle', popular yellow-green Italian-type frying pepper, red when mature, 6-inch fruits are smooth and tapered, 68 days.

'Golden Bell', flesh light to medium green maturing to deep yellow, fruit 4 by 3 inches, a tapered three- or four-lobed bell, 68 days.

'Golden Summer', a four-lobed lime green hybrid, 4½ by 3½ inches, that turns golden at maturity, 67 days (TM).

'Gypsy', 3- by 5½-inch wedge-shaped pepper that changes from light green to golden yellow to red. An All-America Selection hybrid, very tolerant of cool weather, 60 days (TM).

'Keystone Resistant Giant', a popular variety for its tall, sturdy plants and abundant cover with continuous production of large, blocky, four-lobed 3½- by 4½-inch fruits with thick walls, dark green then red, tolerates high heat, 75 to 80 days (TM).

'New Ace', good for northern gardens with a 62-day maturity, three- or four-lobed 4-inch fruits.

'Purple Bell', a hybrid with blocky, four-lobed 3½-inch-square fruits that change in color from green to purple to red, 75 days.

'Sweet Banana' (also called 'Hungarian Sweet Wax'), mild, thick-walled, banana-shaped fruits that are 6 inches long and light yellow turning red, 70 days.

'Whopper Improved', square, four-lobed 4-inch hybrid, 71 days (TM).

'Yolo Wonder', widely adapted market-type hybrid, vigorous 2-foot plants, four-lobed fruits dark green turning red, 4 inches long, matures in 73 days (TM).

**How to use** Most gardeners harvest bell peppers when green, but left on the plant until red, yellow, or purple they become even sweeter and more mellow. To prepare bell peppers, remove the stems, seeds, and pith. Then slice and chop them for use in salads, raw vegetable crudités, soups, stews, omelets, and vegetable casseroles. They can be fried, sautéed, stuffed and baked, or pickled, depending on the type. If a recipe calls for peeled peppers, treat the same way as described under Chiles on the next page.

Chopped bell peppers can be sautéed or butter-steamed in about 5 minutes; they are especially good with onions or mushrooms. Purple bell peppers usually turn green when cooked. Use large, blocky bells such as 'California Wonder' or 'Yolo Wonder' for stuffing with your favorite meat or vegetable mixture.

Since bell peppers have such a short growing season and many are ready at once, dry or freeze some for later use. Cooked stuffed bell peppers freeze well and reheat beautifully.

'Peperoncini' a mild Italian pepper

Chiles

**Pimientos**

The sweetest of all peppers is the pimiento. They are of two types: the cheese, or squash; and the heart shaped. Look for 'Pimiento Select', bright red, heart shaped, smooth, 75 days; and 'Pimiento Perfection', a heart-shaped, thick-walled 2- by 3-inch pepper resistant to tobacco mosaic virus. Dried, pimientos become paprika, the well-known vivid red spice.

**Chiles**

Perhaps the most intensely flavored, most loved, and most confused group of vegetables is chiles. Part of the problem lies in the many varieties: More than 100 have been counted and they all cross-pollinate with great ease. Many varieties have more than one name. If that is not enough, a chile that is mild tasting when grown in the mild-climate conditions of a California coastal valley becomes hot when grown in the more stressful conditions of New Mexico.

Most but not all chiles that are long, tapered, thin, cone shaped, or round are hot. The color of a chile has nothing to do with its spiciness; most chiles are sold when red, but they can be green or yellow as well. When dried, chiles become chili powder or cayenne pepper and are the base for Tabasco and other hot sauces.

In Mexico, all peppers are called chiles and are commonly named by their use, such as *chile para rellenar* (for stuffing), or *huachinango* (to be cooked with red snapper). Some are named for a region, such as Tabasco. Other names are derived from shape (ancho is broad) and color (*guero* is blond). Usually the name changes again when the chile is dried.

Why all the interest in chiles? For one reason, more are produced and consumed than any other spice in the world. Mexican food has become increasingly popular, as have other ethnic dishes that use chiles for seasoning.

**Varieties** The following chiles are the most frequently recommended. Resistance to tobacco mosaic virus is indicated by TM.

'Anaheim TMR 23', a large-fruited cayenne, one of the mildest chiles, green fruits becoming red at full maturity, 6 to 8 inches long, flat, tapering to a point, 77 days (TM).

'Greenleaf Tabasco', especially bred at Auburn University for southern gardeners, gives a heavy yield of bullet-shaped yellow-green fruits that turn red, excellent for sauce, extremely hot, 120 days (TM).

'Hot Portugal', recommended for northern and other short-season areas, sturdy plants, heavy yielding, green turning scarlet, 6 inches long, very hot, 64 days.

'Hungarian Wax' (also called 'Hungarian Yellow Wax', 'Hungarian Hot Wax', and 'Hot Banana'), light yellow then bright red at maturity, fruits tapered, 5 to 8 inches long and about 1½ inches wide, often described as medium hot, 67 days.

'Jalapa', oval tapering to a blunt point, 2½ inches long, hybrid, hot, an improved Jalapeno, 65 days.

Jalapeno, widely adapted, dark green becoming red, 3½ by 1½ inches, tapered to a blunt point, very hot, 72 days.

'Large Cherry', medium green to red, heavy crops, fruit 1 by 1½ inches, good for pickling, hot, 78 days.

'Long Red Cayenne', dark green then red, fruits 5 inches by ¾ inch and often curled or twisted, easily dried, very hot, 73 days.

'Mexi Bell', a mildly hot, three- or four-lobed hybrid bell pepper, an All-America Selection, 70 days (TM).

'Red Cherry Hot', round, 1½ inches across, green turning red, 75 days.

'Red Chili Hot', 2½ inches long, tapered, hot, 80 days.

'Roumanian Wax' (or 'Roumanian Hot'), yellow turning red, medium hot, 2½ by 4 inches and blocky, 65 days.

'Santa Fe Grande', very popular in the Southwest, yellow turning orange-red, 3½- by 1½-inch conical fruit tapers to a point, medium-thick skin, 76 days (TM).

'Serrano Chile', highly recommended in the Southwest, green then red, small 2¼- by ½-inch fruits are slim and club shaped, walls thin, used for pickling and sauces, one of the hottest, 75 days.

Cayenne chiles

'Super Chile' chiles

'Super Cayenne', very hot, long, slim 4-inch peppers turn from green to red, an All-America Selection, 60 days.

'Super Chili', very hot, cone-shaped 2½-inch fruits change from green to orange to red and are borne upright on the plants, an All-America Selection, 70 days.

'Thai Hot', very hot, small, cone-shaped fruits green turning red on very small, compact plants, 64 days.

**How to use** Chiles have an honored place in international cuisine, figuring prominently in Mexican, Indian, African, Spanish, Portuguese, Indonesian, and Korean dishes.

Since the oil from chiles can irritate the eyes and skin, it's a good idea to wear rubber gloves when preparing them; you might also want to hold them under running water. Be sure to remove the stems, seeds, and inner membranes (the hottest part). To peel chiles, blister them under the broiler, then slip them into a brown paper bag, twist closed, and let stand to steam and cool. Remove them one at a time and peel, starting at the stem end.

Chiles can be diced and stored in vinegar to add *picante* to any type of dish. Both the chiles and the vinegar can be used in cooking and salads and atop tacos.

### Drying Peppers

Both bell peppers and chiles can be dried for winter use, the dried chiles usually being called by a name different from the same fresh chile. In Mexico, dried chiles are ground in a mortar for a supply of chili powder. A pinch will add flavor to almost any dish, the flavor intensified by adding increasing amounts of the powder.

Peppers can be dried in a microwave, on a drying rack, or air-dried and strung until needed.

A dried chile important in Mexican cuisine for *mole* is the ancho. When fresh, it is called the poblano and is used in chiles rellenos.

### Japanese Peppers

Japanese peppers are gaining in popularity. An important vegetable in Japan, 'Long Green Fushimi' is a longtime favorite bell pepper. The outstanding chile in Japan is 'Yatsurusa'.

## PERSIAN MELONS

See Melons.

## POTATOES
## *Solanum tuberosum*

Potatoes have been grown and used in temperate climates along the Andes for at least 2,000 years, but it was not until the sixteenth century that Spanish explorers introduced them into Europe.

The English, French, and Germans regarded the potato mainly as a curiosity for more than a century. The Irish were the first to realize its crop potential. They became economically dependent on the potato by 1845, when the fungus disease late blight struck, eventually devastating the crop and causing widespread famine and emigration.

Potatoes became a fairly important crop in North America after many Irish settlers arrived in 1718, and became very significant right after Ireland's famine.

The common white, or Irish, potato is now economically the most important vegetable in the world and the fourth most important plant in terms of food value. In volume and value it is exceeded only by wheat, rice, and corn. Americans eat more potatoes than any other vegetable—about 120 pounds each per year—and for good reason: They are nutritious, economical, versatile, and easily prepared.

Potatoes are annuals and require a frost-free growing season of between 90 and 120 days. The ideal climate has a relatively cool summer, especially at the end of the summer, when the potatoes are maturing. Plant early potatoes just before the last killing frost. Sprouts will develop at low temperatures but may be injured if exposed to frost when above ground. Usually it is safe to plant when the soil is warm enough to be worked, about 45° F.

Main fall crops are usually planted 120 days before the first killing frost in fall. In short-season areas, plant as early as you can to harvest a late crop.

Harvesting new potatoes

Potato plant in bloom

**Soil preparation** Potatoes need a rich, loose, slightly acid soil. Use plenty of organic matter and add 5-10-10 or a similar low-nitrogen fertilizer, about 10 pounds per 100 feet of row, or as instructed on the fertilizer label.

The recommended soil pH for potatoes is 4.8 to 5.4. If the soil is not acid enough, scab disease, which causes brown corky tissue on the potato surface, may be a problem. Lime the soil for potatoes only if a soil test shows a pH below 4.8.

**How to plant** Although there are a few varieties of potatoes that can be grown from true seed, they are usually grown from seed potatoes, a small piece of potato with at least one growing eye, which will sprout into a new plant. Buy certified disease-free seed potatoes. This is important because potatoes can host many diseases that reduce growth but are otherwise undetectable in most gardens. It makes good sense to try to avoid bringing potato diseases into your garden.

Good-sized seed pieces increase the chances of a good yield. Cut them about 1½ inches square, making sure that each has at least one good eye, and cut a week before planting to allow cut surfaces to heal slightly. Some growers dip cut pieces into a dilute bleach solution or commercial fungicide to prevent rot. You can also plant small potatoes whole and avoid the risk of rot altogether.

Set the pieces, cut side down, eye up, about 4 inches deep and 12 inches apart in rows 2 to 3 feet apart. (Ten to 12 pounds of potatoes will plant 100 feet of row and yield 1 to 2 bushels at harvest.)

Potatoes form not on the roots but on the stems rising from the seed that are above the roots. Sprouts usually appear after 2 to 3 weeks, unless they were ¼ inch or longer when planted or under less than 2 inches of soil.

If the plants have been spaced correctly, the foliage will shade and cool the soil as the tubers mature, preventing high temperatures from damaging them. To further cool the soil, mulch 6 inches deep with a loose organic material.

When plants are 5 to 6 inches high, hill up the mulch and soil around the growing stems. Potatoes exposed to light turn green, an effect associated with the naturally occurring poison solanine. (Small amounts of green tissue can be scraped away, but excessively green potatoes should be discarded.)

**Fertilizer and water** To fertilize when planting, place seed pieces in the center of a 6-inch-wide trench and work the fertilizer in at the edges with a cultivator. Do not let the fertilizer touch the seed pieces. Too much nitrogen fertilizer may cause excessive leaf growth at the expense of the tubers. Fertilize, but not too much.

Potatoes need a steady moisture supply. If the soil dries out after the tubers begin to form, growth stops. It starts again as the soil is watered. The result of this stop-and-start growth is misshapen, knobby, split, or hollow tubers. Try to keep the soil moist to a 1-foot depth through the growing season. Barring rain, that normally means one heavy watering weekly.

**Harvesting** Pick new potatoes as soon as the tops flower. New potatoes are not a variety but are simply any potatoes harvested before full maturity. They are smaller and more tender but will not store. If the soil is loose, simply reach in; otherwise, gently uproot the plant to check its progress. Harvest some potatoes from each plant for new potatoes and leave the rest for harvest when the potatoes are completely mature.

Potatoes headed for winter storage need to mature fully in the soil. For full-sized tubers, wait until the vines yellow or die back. To avoid bruising the potatoes, loosen the soil with a rake or a spading fork first, then dig the potatoes. Store them in the dark for a week or so at 70° F to heal bruises and condition them. Then store them at between 35° and 40° F, keeping the humidity high.

**Beginners' mistakes** You can try growing plants from grocery store potatoes, but they frequently carry diseases and may have been treated to prevent sprouting.

Potatoes grown in a wire cage

Round red potatoes

Some other common mistakes are as follows.

- Overfertilizing before tubers are formed.
- Ignoring the best planting dates.
- Allowing tubers to receive sunlight, making them green and inedible.
- Damaging the shallow stems on which potatoes form during the cultivation that is necessary to reduce weed competition.

**Special handling** An old-time method of growing potatoes is to set seed pieces about 3 inches deep in a trench. As the stems grow, cover them with straw, leaves, pine needles, or any similar material. (If wind is a problem, anchor the straw with some soil.) Potatoes then can be picked simply by pulling back the straw.

Because a 100-foot row is needed to grow the 1 to 2 bushels that most families need, potatoes are usually not considered a vegetable for small gardens. But their growth habit makes possible some interesting experiments.

Potato tubers grown in any loose material such as straw will be cleaner, better formed, and easier to harvest than those grown in soil. They can be grown in plastic bags, adding loose soil to fill the bag as they grow. They can also be grown in containers.

### Potato Varieties

There are many varieties of potatoes but only four basic types: russets, round reds, round whites, and long whites. Although most potatoes have white flesh, there are some with yellow, purple, and even bluish tinged flesh. Varieties are sometimes classified by the use for the potato—baking or boiling. The difference between the two is primarily in the flesh, with boiling potatoes being moist and baking potatoes being dry and mealy. The skin on boiling potatoes is usually thinner than that on baking potatoes. Baking potatoes, because of their thicker skins, store better; boiling potatoes are better for making french fries.

In the following varieties disease resistance is indicated as follows: late blight (LB), scab (S).

**Early** The following varieties mature in 90 to 110 days.

'Haig', a white potato with somewhat flaky skin (S).

'Irish Cobbler', an oblong white potato of wide adaptation, one of the earliest varieties of unknown origin, susceptible to scab.

'Norgold Russet', an oblong to long white with shallow eyes and netted skin, not as early as 'Norland', excellent for baking and boiling but does not store well (S).

'Norland', widely adapted red potato, one of the most favored by home gardeners, very early, medium-sized oblong tubers that can be used many ways but generally recommended for boiling (S).

**Midseason** The following varieties will mature in 100 to 120 days.

'Chieftain', attractive, smooth red-skinned tuber, seed pieces should be cut 10 to 12 days before planting if possible to allow for healing cut surfaces, good quality, many kitchen uses, some resistance to several diseases.

'Red La Soda', a good round red potato, frequently recommended in the South, one of the best adapted in Wyoming, good producer, good for boiling and potato salad, does not store well.

'Superior', white, fairly early maturity, medium sized, roundish, good yield, good boiling potato sometimes used for potato chips (S).

'Viking', large red, excellent cooking qualities; for the best yield they should be planted close in the row to compensate for a tendency to produce a few large tubers.

**Late maturing** The following varieties mature in 110 to 140 days.

'Butte', a russet-skinned hybrid very similar to its grandparent 'Russet Burbank', promising the same high kitchen quality but with more protein and vitamin C; also has shown improved yields and resistance to scab disease.

'Katahdin', white skinned, widely adapted, large round to oblong tubers that are excellent for baking.

'Kennebec', white skinned, block shaped, excellent eating quality, one of the best for baking, frying, and hash browns, stores moderately well (LB, S).

'Red Pontiac', red skinned, widely adapted, oval tubers that may become too big with abundant rainfall, fair table quality, very good storage

'Katahdin' potatoes

'Guilio' radicchio

quality, generally used for boiling.

'Russet Burbank' (or 'Idaho Baker' or 'Idaho Russet'), the most important fall-harvested potato in the United States and western Canada, large tubers are long and cylindrical, a very long growing season; excellent for baking, frying, hash browns, storing (S).

**How to use** Contrary to some claims, potatoes are high in nutrition—an excellent source of protein, minerals, and vitamin C—and relatively low in calories.

For the most food value, leave the skin on whenever possible, even when frying; if you must peel potatoes, keep parings thin. If not cooked immediately after paring, potatoes should be covered with cold water to prevent darkening.

To prevent potatoes for a stew from becoming soggy, boil them separately, peel, and add during the last few minutes of cooking. For light, creamy mashed potatoes, add hot milk instead of cold.

## PUMPKINS

See Squashes, Pumpkins, and Gourds.

## RADICCHIO
### *Cichorium intybus*

Radicchio is a type of chicory that develops a small, tight head of red to magenta leaves with white veins. Mature plants range in size from a golf ball to a softball.

Radicchio needs a long, cool growing season, so plant seeds outdoors in midspring in the north and in late summer in the south. Thin plants to 8 to 10 inches apart. Older varieties need to be cut back in early fall to force the heads to develop in 4 to 6 weeks, but this is not necessary with newer varieties. Harvest heads of spring-planted radicchio in fall, and late-summer-planted radicchio in late winter or early spring.

**Varieties** 'Guilio' produces deep burgundy heads without cutting back, 80 days; 'Marina', sown in late spring, forms heads in fall without cutting back, 110 days; 'Red Verona' must be cut back to force heads, 85 days.

**How to use** Radicchio has a sharp flavor and is used with other salad greens for its pungency and color. Its raw leaves make an attractive garnish. It may also be steamed or sautéed.

## RADISHES

See Root Crops.

## RHUBARB
### *Rheum rhabarbarum*

This hardy perennial is grown for its edible leafstalks. The leaves contain poisonous quantities of oxalic acid.

Rhubarb varieties adapted to the northern United States and Canada require 2 months of temperatures around freezing to break their rest period; for quality and yield, rhubarb also needs a long, cool spring. It does not grow well where summer temperatures exceed 90° F. Plant in early spring as soon as the soil can be worked.

Start rhubarb from root divisions (rooted crowns) with 1 to 3 buds (eyes). Rhubarb can be grown from seed, but the results are variable and not the quality of selected crowns. Root divisions offered by local garden centers are picked for quality and climate adaptability.

Plant in a trench 12 to 18 inches deep and filled with a rich soil mix to within 2 to 3 inches of the top. Set crowns about 2 inches below the soil surface. Do not allow them to dry out before planting.

Rhubarb needs space to grow—3 feet between plants. But 3 or 4 plants will supply all the average family can use. Locate the plants out of the way of regular gardening operations.

Rhubarb will produce for 4 to 6 years. The first year after planting, allow all stalks to grow and do not harvest. The second year harvest only for 1 to 2 weeks. After that, you can have fresh rhubarb for 8 weeks or more, but do not remove more than half the stalks at any one time.

Divide rhubarb every 4 years or when the stalks start to become thin. Fertilize plants in early spring and again after harvesting, and apply winter protection where temperatures drop below 10° F.

Rhubarb

Beets

**Varieties** The popular varieties in cold-winter areas are 'McDonald's', with bright red stalks; 'Valentine', with deep red stalks that retain their color when cooked and are the sweetest; and 'Victoria', with green stalks sometimes tinged in pink, the most reliable variety for starting from seed. In mild-winter areas, use red-stalked 'Giant Cherry', which needs less cold than other varieties to break its winter rest period.

**How to use** If the stalks are young and tender, there is no need to peel them before cooking; just wash them and cut into 1-inch chunks. Older stalks, however, may need to be peeled and de-strung as you would celery.

Always cook rhubarb. Steam it in a double boiler or stew it like applesauce, adding sugar to taste near the end of the cooking time. Many cooks spoil rhubarb by sweetening it too much, which destroys its natural tartness. Rhubarb is often combined with strawberries in salads and fruit compotes. Try it in a molded gelatin salad with cream cheese, chopped celery, and nuts. Alone or with berries, it makes good jelly and jam.

Rhubarb is best known in an American original—rhubarb pie—which comes in many forms: double crust, deep dish, meringue, and even custard. Rhubarb is also popular in cakes and puddings; spiked with port or brandy, it becomes a delicious topping for ice cream.

## ROOT CROPS

### Beets
### *Beta vulgaris*, *Crassa* Group

The original home of the beet was around the Mediterranean, where it first occurred as a leafy plant without enlarged roots. Improved types of these early beets are now grown as Swiss chard. Large-rooted beets were first noted in literature around 1550 in Germany, but only one variety was listed in the United States in 1806.

Beets are round or tapered roots; although usually red to purple, there are varieties with white or yellow roots as well. Beets are technically biennials but are grown as annuals because the roots become tough and stringy in the second year.

Although beets prefer cool weather, they tolerate a wide range of conditions. They can be planted early and will withstand a light frost, but additional plantings can be made for harvest over a long period. In very hot weather special attention to watering and mulching may be needed.

Beets do not transplant well. Sow beet seeds directly in rows a foot or more apart and then thin to 2 inches. Some thinning can be postponed until the extra plants are large enough for edible greens and roots. Unless you use a monogerm (single-seeded) variety, each beet seed (actually a seed ball) will produce two to five seedlings in a tight clump, so some thinning should be done early.

**Beginners' mistakes**
The most common problems are overplanting and underthinning.

Stringy and tough beets are the result of a lack of moisture, which may be caused by drought or by overcompetition from other beets or weeds. As with most vegetables, beets must be grown at full speed, without a letup. The soil must be rich, loose, and well drained for the best root production.

Harvest beets when they are 2 to 3 inches across; gently poke your finger into the soil around the beet to check its size before pulling.

**Varieties** The choice of varieties for garden use is not very critical. (Downy mildew resistance is needed in certain areas.) All varieties can serve both for roots and greens, but if greens are needed to any extent it would be better to plant chard or a variety of beet selected for that use.

'Burpee's Goldenbeet', unusual golden yellow root, good quality, may average higher in sugar, color does not bleed out in cooking, excellent for greens, low germination rate, 55 days.

'Cylindra', long cylindrical dark red root gives many uniform slices, up to 8 inches long, 1¾ inches in diameter, 60 days to maturity.

'Detroit Dark Red', dark color, neat globe shape, downy

Beets and turnips

Assorted carrots

mildew resistant strains available, 60 days.

'Early Wonder', semiglobe shape, deep red flesh with lighter zones, 55 days.

'Green Top Bunching', round, blood red beets and delicious bright green foliage, 58 days to maturity.

'Little Ball', a baby beet that grows fast and stays small; serve whole or can or pickle, 56 days.

'Lutz Green Leaf' (also known as 'Winter Keeper' because it stores well), top-shaped roots with dark red flesh with lighter zones, glossy green foliage with pink midribs make excellent greens, 80 days to maturity.

'Mono-King Explorer', deep red, monogerm type, less thinning required, 50 days.

'Perfected Detroit Dark Red', medium-sized dark red globe, improved color, shape, and uniformity compared to 'Detroit Dark Red', 58 days.

'Red Ace', globe-shaped hybrid with smooth dark red flesh, 53 days.

'Red Ball', globe shaped, smooth, fiberless dark red flesh, good for areas with a cold spring, 60 days.

'Ruby Queen', globe shaped, ringless deep red flesh, 52 days.

**How to use** The beauty of a beet is that you can eat all of it. Cook the nutritious tops as you would other greens, then toss with a mixture of butter and bread crumbs, or garnish with diced hard-cooked egg. Drop whole beets into boiling water; when tender, the skins will slip off easily. Cook beets with an inch of leafstalk attached to keep them from bleeding. A bit of lemon juice or vinegar in the water further stabilizes the red pigment and adds flavor. Shred and cook in butter; or slice and serve hot with butter, an orange sauce, or sour cream.

Beets are the main ingredient in borscht, a soup usually served cold and garnished with sour cream. Beets are excellent when pickled; when cooked and chilled they can be added to salads.

Tiny beets, the thinnings from the garden, can be cooked intact—tops and roots together. Steam them in a covered heavy pan with lemon juice, a chopped green onion, several tablespoons of oil, and your choice of seasonings.

### Carrots
***Daucus carota* var. *sativus***

Carrots as we know them originated from forms grown around the Mediterranean. By the thirteenth century carrots were well established as a food in Europe and came with the first settlers to America, where Indians soon took up their culture.

Carrots are adaptable, tolerant of mismanagement, and unequaled for storing well over a long period, using nothing more complicated than the soil in which they are grown. Carrots can be left in the ground over the winter, but they are usually less tender and tasty than if they are pulled in the fall.

Carrots are grown from seed directly sown into the garden. Like most root crops, they do not transplant well. The best planting times are early spring and early summer. Each planting will mature in 60 to 85 days and can be harvested over a 2- to 4-month period. Small plantings every 3 weeks will ensure a continuous harvest. Sow seeds ½ inch deep; germination takes several weeks. Thin plants to 3 inches apart and use the thinnings in salads, stews, and soups.

Soil for carrots must be loose and very rich or the roots will be deformed. Manure, unless very well rotted, causes roughness and branching. Soil that is allowed to dry out causes carrots to split.

Harvest carrots when they are 2 inches or less across; the smaller the tastier.

**Varieties** Carrots differ mainly in their size and shape. Some are long, slender, and tapered; others are short, thick, or even round. The short to medium or very short types are better adapted to heavy or rough soils than the long types. They are also easier to dig. The tiny finger carrots are often the sweetest and grow well in containers.

**Long** The following varieties range from 8 to 10 inches long. 'A Plus', tasty deep orange carrot that supplies almost twice as much vitamin A as other carrots, 71 days; 'Gold Pak', particularly deep color, 8 to

Harvesting carrots

Harvesting parsnips

9 inches long, 76 days; 'Imperator', the standard market carrot, good quality, 8 to 9 inches long, 75 days.

**Medium** The following varieties range from 6 to 8 inches long when mature.

'Danvers Half Long', almost perfectly cylindrical 6- to 7-inch roots, 1½ inches thick, 70 days; 'Nantes', favored for flavor and tenderness but tends to crack in wet fall weather, 68 days; 'Royal Chantenay', improved strain of 'Red-Cored Chantenay', 6 to 7 inches long, slightly tapered, holds well in the soil, 70 days; 'Touchon', French import, long and slim, fine textured, practically coreless and excellent for juicing, 75 days.

**Short** The following varieties range from 3 to 6 inches long when mature.

'Ox Heart', plump, recommended for heavy soils, 75 days; 'Short 'n Sweet', 3- to 4-inch bright orange carrot, good in heavy soils and containers, 68 days.

**Finger carrots** These are ideal for containers. A lightweight soil mix is important, since their tops break off easily when being pulled from heavy soil. Roots are tender, sweet, and just right for eating whole or pickling. Three popular varieties are 'Lady Finger', tender, sweet, 5-inch carrots, 65 days; 'Little Finger', 3½ inches long and very tender, 65 days; and 'Tiny Sweet', 3 inches long, 65 days.

**Ball-shaped carrots** More round than long, these are excellent for container growing, serving on relish trays, and canning in whole-pack jars. Look for 'Planet', a very sweet carrot that can also be grown in a cold frame or in shallow or rocky soil.

**How to use** To make the most of this nutritious vegetable, don't pare or even scrape carrots; just scrub them well. For a relish tray, soak them in ice water for extra crispness. Shredded carrots, raisins, pineapple chunks, and mayonnaise make an excellent summer salad; shredded carrots are almost as important to coleslaw as cabbage.

To cook, boil carrots quickly in salted water. Serve with butter and a sprinkling of parsley and chives. Added to slow-cooking stews, they improve as they take on the flavor of the meat. Cook them in a cream sauce seasoned with tarragon, nutmeg, and dill weed; or steam with mint leaves for a superb accompaniment to lamb. Bake carrots in bread crumbs, or dip in batter and deep-fry. Try them baked in carrot bread, pie, or cake.

### Parsnips
### *Pastinaca sativa*

Parsnips are native to the Mediterranean. They were common in Europe by the sixteenth century, and the early colonists brought them to North America.

Parsnips grown from seed produce 1- to 1½-foot white to cream-colored roots in 100 to 120 days. Parsnip seeds cannot be stored from one year to the next. Sow seeds ¼ to ½ inch deep directly into the garden in early spring. Sow heavily, since the germination rate is low, and thin to 5 inches apart after they are 1 inch high.

Parsnips need light, loose soil, not only to grow undistorted roots but so they can be harvested intact. They benefit from frequent fertilizing and weeding.

Parsnips can be left in the ground all winter. Mulch them heavily to harvest throughout the winter, or pull them before growth begins in the spring. They need winter cold near the freezing point to change their starch to sugar and develop the sweet, nutlike flavor for which they're famous. They can also be dug in the late fall and stored in moist sand. Although they can withstand alternate freezing and thawing in the ground, being frozen after harvest will damage them.

**Varieties** The standard varieties are 'All American', very tapered and shorter and wider than other varieties, 110 days; 'Harris Model', with smooth, very white roots, 110 days; and 'Hollow Crown', a long, smooth variety, 100 days.

**How to use** Parsnips have a distinctive flavor and a small but enthusiastic following. To prepare, steam them in their skins, then peel and slice lengthwise. If a large core has developed, cut it out. Pan glaze with butter, a touch of brown sugar, and nutmeg for a delicious alternative to candied sweet potatoes. Or brown them and dress with chopped walnuts and cream sherry.

Assorted radishes

'Round Black Spanish' and 'China Rose' winter radishes

Add parsnips to a roast during the last half hour of cooking and baste frequently with pan juices, or slice them into a soup or stew.

### Radishes
### *Raphanus sativus*

Give a youngster a package of radish seeds and say "go plant" and you'll have radishes. But to grow crisp, mild, nonpithy radishes, you must adhere to the fundamentals of fertilizing and watering.

Fertilizers must be worked into the soil before planting to be immediately available to the young seedlings, since radishes must grow quickly for the best flavor. A second application of liquid fertilizer 2 weeks after germination will benefit the crop. The soil should be loose, rich, and moist. Spring radishes mature in 3 to 4 weeks, so there's little time to correct mistakes.

Spring radishes are grown from seed directly sown into the garden in early spring as soon as the soil can be worked. Sow every 7 to 10 days until early summer, then start again in late summer. Winter radishes should be sown so they reach maturity during the fall. Thin seedlings to 1 to 2 inches apart very soon after emergence to reduce competition, because roots begin to expand when only 2 weeks old. Scatter seeds spaced out in a 3- or 4-inch-wide row to reduce the need for thinning.

Radishes are cold hardy but do not withstand heat. In the South they grow well in fall through spring. In the North they are usually grown in spring and fall. Where summers are cool, they can be harvested all summer for salads and garnishes.

**Varieties** Radishes have either round or elongated roots, which may be white, red, or bicolored. The radish varieties listed here are divided into two groups—spring and winter—even though the term *spring radish* is misleading. Spring radishes can be grown throughout the season in cooler areas and in all but the hottest months in warmer areas.

Spring radishes come in many shapes and sizes.

'Burpee White', round, white skin and flesh, harvestable at ¼ to 1 inch, 25 days.

'Champion', round to oval, bright red, lasts in the garden to 2-inch diameter without becoming pithy, 28 days.

'Cherry Belle', quite round, red with white flesh, cherry sized, an All-America Selection, 22 days.

'Crimson Giant', globe shape, 1½ inches wide, should be thinned to 1½ inches, crisp and mild, deep red skin, white flesh, 29 days.

'Early Scarlet Globe', olive shaped, bright red skin, fast growing, 24 days.

'Easter Egg', hybrid, white flesh with multicolored red, purple, pink, violet, and white skin, 25 days.

'French Breakfast', oblong, red with a white tip, white flesh, 24 days.

'White Icicle', white skin and flesh, 5 inches, tapered, 28 days to maturity.

Winter radishes (*R. sativus* var. *longipinnatus*) are so called because when they are planted in the spring they tend to flower before harvestable roots can develop. They need the decreasing temperatures and day length of fall and winter to discourage this flowering.

The winter radishes include the Asian, or daikon, types listed below. They are slower growing, much larger, more pungent, and longer keeping than spring types, and can be stored over winter like other root crops.

'Celestial' ('White Chinese'), pure white skin and flesh, mild, 6 to 8 inches long, 60 days.

'China Rose', long, hot, 52 days to maturity.

'Miyashige', 15 inches long, 2 inches wide, 60 days.

'Round Black Spanish', globe-shaped roots to 4 inches across, black skin, crisp white flesh, 55 days.

'Sakurajima Mammoth', giant size—to 70 pounds, 70 days.

'Takinashi', white, brittle, 12 inches long, 65 days.

**How to use** Both red and white varieties of spring radishes are most popular eaten raw, either alone or in salads. Scrub them, chill, and serve with a bowl of salt or your favorite dip.

Any radish can be cooked, but it's more common with the large winter varieties, such as the daikon. Steam them with sautéed green onions, add salt

Salsify foliage

'Sandwich Island Mammoth' salsify

and pepper, and serve with a cream or cheese sauce.

Daikon radishes are used in Asian dishes. Shredded daikon is the classic Japanese accompaniment to sashimi.

### Salsify
### *Tragopogon porrifolius*

The flavor of the tapered 8-inch-long salsify root earned it the names vegetable oyster and oyster plant. It is a biennial but can be grown as an annual, growing 2 to 3 feet high with grassy leaves. Long-stemmed purplish flower heads appear the second year if the roots are not harvested the first fall.

Like carrots and parsnips, salsify grows the best roots in a deep, rich, loose, crumbly soil. Fresh manure should not be used, since it causes the roots to split.

Plant as soon as the ground can be worked in spring. Sow seed ½ inch deep in rows 16 to 18 inches apart. When seedlings are 2 inches high, thin to 2 to 3 inches apart. Keep the soil evenly moist during the growing period.

Harvest the roots in fall before the ground freezes. The roots can also be left in the ground all winter and dug in the spring. Mulch thickly in cold-winter areas.

**Varieties** The standard variety of salsify is 'Sandwich Island Mammoth', 120 days. Black salsify, or *Scorzonera*, is actually a different genus, but it is used and grown similarly. It has black roots with white flesh and, in the second year, dandelionlike flowers on 2½-foot stems.

**How to use** Salsify, with its distinct oysterlike flavor, can be served as a main dish or as an accompaniment to any meat. To prepare, scrape the roots, cut into 2- to 3-inch lengths, and place in cold water containing a few drops of lemon juice or cider vinegar to prevent discoloration. Serve either raw or cooked.

To make mock oysters, boil salsify and drain, then dip in egg, roll in flour or bread crumbs, and sauté in butter until tender.

### Turnips (*Brassica rapa, Rapifera* Group) and Rutabagas (*Brassica napus, Napobrassica* Group)

Turnips originated in western Asia and the Mediterranean in prehistoric times. Rutabagas are more recent, apparently originating in the Middle Ages from a cross of turnip and cabbage. Rutabagas are sometimes called Swedish turnips.

Although these two crops are often considered together, as here, there are distinct differences. There are white and yellow forms of each; however, most turnips are white fleshed, most rutabagas yellow. Turnips have rough, hairy leaves, are fast growing, and the roots become pithy in a short time. Rutabagas have smooth, waxy leaves, emerge and develop much more slowly, are more solid, and have a long storage life. Rutabaga roots are larger, sweeter, and much higher than turnips in vitamin A and most other nutrients. The tops of both vegetables are outstanding sources of vitamins A and C.

Any turnip can be grown for greens. There are turnip varieties bred especially for greens that do not produce edible roots.

Turnips and rutabagas prefer cool weather, and a light frost at maturity improves their flavor. Both spring and fall plantings work well with turnips, since they mature in 60 days or less. Plant in early spring as soon as the ground can be worked for a late spring crop, or in late summer for a fall crop. In warmer areas, fall and winter crops are more successful, because they mature at the end of the cool season.

The turnip's short season permits it to be grown at some time everywhere in the United States. Rutabagas, which take more than 90 days to mature, are grown in the South in fall and winter, and in northern areas in late summer and fall when temperatures average 75° F or less.

Direct sow about ½ inch deep in rows as close as 15 to 18 inches apart. Thin turnips to 3 to 5 inches apart; rutabagas to 6 to 8 inches apart. Keep them well watered and growing fast.

Turnip ready for harvest

'Tokyo Cross' turnips

General soil and nutrient requirements are about the same as for beets (see page 117), although turnips and rutabagas may need slightly less nitrogen. Seedlings usually grow quickly and easily, so the seedbed need not be extremely fine.

Harvest turnips when they are 2 to 3 inches across; if grown as a spring crop, harvest before 80° F weather. Turnip greens are harvested when young, after about half the time needed to produce a mature root crop. Greens can be harvested from turnips being grown for roots if the growing tip is not damaged; the plants will produce new greens and the roots will continue to grow. Harvest rutabagas when 3 to 4 inches across; they can be left in the ground until it freezes.

**Turnip Varieties**

'All Top', grown for its tops, has thick, smooth dark green foliage, 50 days.

'Early Purple-Top Milan', flattened 3- to 4-inch roots, 45 days to maturity.

'Just Right', white hybrid for greens or roots, 37 days.

'Purple-Top White Globe', the standard variety for roots, 58 days.

'Tokyo Cross', 2-inch pure white roots but can grow larger without becoming pithy, All-America Selection, 35 days.

**Rutabaga Varieties**

'American Purple Top', an old-timer, roots are purple above ground and light yellow below, yellow flesh, 88 days.

'Macomber', very sweet and fine grained, stores well, 92 days.

**How to use** Turnip greens go well in salads when young and tender. They can also be steamed or boiled, and are especially good cooked with other greens such as mustard and chard.

Turnips should be peeled before using. Slice and serve raw, or boil and serve with parsley and lemon butter. Baked, mashed, scalloped, or country-fried turnips are excellent with pork and game. Mashed turnips can be combined with mashed potatoes. Turnips can also be used in soufflés, soups, stews, and casseroles. Asian cooks use them in stir-fried dishes and for pickling.

Peel rutabagas before cooking; if cooking in water, add a teaspoon of sugar to improve their flavor. Rutabagas can be baked, french fried, glazed, boiled, and mashed. They are good creamed with minced onion and seasoned with Worcestershire sauce, or as the prime ingredient in a vegetable soufflé.

Try seasoning rutabagas with nutmeg, fresh mint, dill, or basil. The tops can be cooked just like turnip greens.

## SALSIFY

See Root Crops.

## SHALLOTS

See Onion Family.

## SHUNGIKU
## *Chrysanthemum coronarium*

*Shungiku* is often referred to as chop suey greens, a more general name that also includes other plants of similar use. Another name is garland chrysanthemum, which is somewhat more appropriate, because the plant is an edible kind of chrysanthemum. The leaves are similar to those of horticultural chrysanthemums, and the plant has bright yellow, daisylike flowers.

Shungiku is grown as an annual. Plant seed in early spring in rows 1 to 1½ feet apart. Or plant in solid beds as with carrots or lettuce. Once plants are 5 inches high, you can begin to harvest, either by pulling the whole plant to thin at the same time or by just removing some leaves. Grow some plants to maturity for a crop of edible flowers.

**How to use** The luxuriant young growth made in the cooler parts of the year are best. The flavor is strong but not sharp or bitter. Shungiku can be used fresh in salads or as an addition to stir-fried vegetables. The flowers can be added to salads or used as a garnish.

## SORREL
## *Rumex acetosa*

Sorrel, or sour grass, has tart, lemon-flavored, arrow-shaped leaves that are high in vitamin C. Plants mature in 100

Spinach patch

'America' spinach

days, from seed to harvesting the leaves from the large, dense clumps.

Sow seed outdoors in early spring, ⅛ inch deep and 4 inches apart, and thin plants to 10 inches apart. Sorrel can be grown as a perennial; if so, divide plants every 2 to 3 years. It tolerates summer heat, but its flavor is less bitter when it matures during cooler weather.

Cut leaves as needed, or remove the entire plant from the ground. Grown as a perennial, it can be harvested by cutting off the leaves at ground level during the spring and then reharvesting in fall.

**How to use** Sorrel can be used fresh in salads, usually combined with chard, spinach, or other greens, or added to stews, sauces, and soups. Cook like spinach and season with mustard and tarragon. Puréed sorrel is a good side dish with fish, goose, and pork.

## SOYBEANS

See Beans.

## SPINACH
## *Spinacia oleracea*

Spinach originally came from what is now Iran and adjacent areas. It spread to China by A.D. 647 and Spain by A.D. 1100, and came to North America with the first colonists.

Growing spinach is a problem for many home gardeners. The biggest obstacle is its tendency to bolt, which stops the production of usable foliage.

Bolting in spinach is controlled by day length and is influenced by temperatures and variety selection. Long days hasten flowering, an effect increased by exposure to low temperatures during early growth and high temperatures in the later stage.

These circumstances make spring culture of a variety that is susceptible to bolting almost sure to fail; therefore, bolt-resistant (or long-standing) varieties must be used in spring. Some of the quick-bolting varieties that are otherwise good can be used in the fall, and in mild areas during the winter.

Make spring plantings in northern areas as early as possible, sowing seeds every 2 weeks until 6 weeks before daytime temperatures reach 75° F. Make fall plantings about a month before the average date of the first frost. In mild-winter areas plant anytime from about October 1 to March 1.

Refrigerating seeds for 1 week before sowing will hasten germination if the soil is above 40° F. Sow seeds ½ inch deep and 2 inches apart in rows 15 to 18 inches apart; thin to 4 to 5 inches apart and use the thinnings in soups and stews. Spinach likes fertile neutral, moist soil.

Harvest spinach when leaves are 6 to 8 inches long and before the plants start to bolt. Cut off the outer leaves or the entire plant.

**Varieties** Besides the tendency to bolt, you may need to consider varietal resistance to three important diseases: downy mildew, or blue mold; mosaic virus; and spinach blight, or yellows. Varieties also differ by having either savoyed (heavily crinkled) or smooth leaves. Savoy types are harder to clean but are dark green, thick, and usually preferred. Resistance to disease in the following varieties is indicated as blight (B), downy mildew (DM), and mosaic (M).

Long-standing varieties include: 'America', a savoy type, dark green, 50 days; 'Bloomsdale Long Standing', a savoy, dark green, slow to bolt, 48 days; 'Melody', a medium savoy hybrid, All-America Selection, 42 days (DM, M); and 'Vienna', a deeply savoyed hybrid with thick leaves, matures in 40 days (DM, B).

Varieties for fall and winter only include: 'Dixie Market', 40 days (DM); 'Hybrid No. 7', 42 days (B, DM); and 'Virginia Savoy', 42 days.

**Summer spinach** In the summer months when cool-season spinach fails the gardener, warm-season tropicals are available as substitutes. They are as rich in vitamins as true spinach and as comparable in flavor.

**Malabar spinach (*Basella alba*)** When the weather warms, this attractive, glossy-leaved vine grows rapidly to produce edible shoots in 70 days. Start seeds indoors and transplant seedlings to the garden in late spring. Train

New Zealand spinach

Assorted winter squash varieties

plants against a fence, wall, or trellis. Young leaves and growing tips can be cut throughout the summer. Use cooked or fresh in salads. 'Red Stem' is an attractive variety with red stems and dark green leaves.

**New Zealand spinach (*Tetragonia tetragonioides*)** This low-growing, groundcover type of plant spreads to 3 to 4 feet across. The tender young stems and leaves can be cut repeatedly throughout the summer. The seeds are actually bundles of seeds, like beet seeds, and are slow to germinate. Start indoors in peat pots and set out after the last frost in spring.

**Tampala (*Amaranthus tricolor*)** The heart-shaped leaves are 4 inches long and green, tipped with orange, red, or purple. Sow seed directly into the garden after the frost danger has passed, making successive sowings every 2 weeks. Greens are ready for harvest in 70 days.

**How to use** Spinach may be served raw in salads or alone with dressing and a garnish of crumbled bacon and diced hard-cooked egg. For a traditional spinach salad, wilt the leaves with a hot dressing of bacon fat, then add vinegar, mustard, honey, chopped green onion if you want, and crumbled bacon.

The trick to preparing good cooked spinach is quick cooking in as little water as possible; what clings to the leaves after washing is enough. Cover the pot and cook until tender (5 to 10 minutes). Drain and season with butter, salt and pepper, and, if desired, a touch of lemon juice.

Serve poached eggs on a bed of spinach, or add spinach to omelets. Try it in a soufflé, or use it creamed as a crêpe filling or as a filling for bell peppers or tomatoes.

Prepare summer spinach like true spinach, and enjoy the hardy greens throughout the summer. They're good cooked, or served raw in salads, and are popular in some Asian dishes.

## SQUASHES, PUMPKINS, AND GOURDS

These plants are members of the gourd family, or *Cucurbitaceae* to botanists. All are native to the Americas. Most pumpkins and squashes originated in Mexico and Central America and were used all over North America by the Indians. Most winter squashes originated in or near the Andes in northern Argentina.

Squashes are divided into two types—summer squash and winter squash. All summer squash is *Cucurbita pepo;* it is a large, bushy plant whose fruit is harvested and eaten when immature, when the rind is soft. Winter squash is primarily a vining plant whose fruits are harvested when mature, when the rind is hard; they are so named because they store well over the winter. Acorn and spaghetti squash are *Cucurbita pepo;* butternut squash is *C. moschata;* hubbard and turban squash are *C. maxima.* Pumpkins fall into different *Cucurbita* species, and gourds fall into several different genera. (The entire family and its relationships are outlined in the chart on page 128.)

**How to grow** Squashes, pumpkins, and gourds, except for the bush types, are space users, not practical for minispace gardeners lacking ways to use vertical space, such as training up a fence or trellis. Even a compost pile can serve as a place for a vine to ramble. They are sometimes also grown with corn, but in that case should be spaced some distance apart.

On the ground, vining types need 10 feet or more between rows but can be grown in less space by training or pruning. Long runners may be cut off after some fruit sets if a good supply of leaves remains to feed the fruit.

Squashes, pumpkins, and gourds are warm-season crops that are heat resistant. Plant them when the soil is thoroughly warm, usually a week after the average last frost. Direct seeding is best, but if the season is too short to mature a direct-sown crop, use transplants from nurseries or start your own in individual pots 4 to 5 weeks before the time to plant out safely. For the best results use hot caps or row covers to reduce transplant shock.

Leave 2 to 4 feet between plants, depending upon the

Straightneck squash

Zucchini

vigor of the variety. Bush types do best in rows 5 to 6 feet apart, but they can be grown as little as 16 to 24 inches apart in the row.

Fertilizing and watering requirements are the same as for cucumbers and melons (see pages 89 and 99). Squashes, pumpkins, and gourds need generous amounts of organic matter in the soil, and 2 pounds of 5-10-10 fertilizer to 50 feet of row. Watering should be slow and deep. Leaves may wilt during midday but should pick up again as the day cools.

Don't worry when the first blossoms fail to set fruit. Some female flowers will bloom before there are male flowers for pollen, so they will dry up or produce small fruits that abort and rot. This is natural behavior, not a disease. The same thing happens when a large load of fruit is set and the plant is using all of its resources to develop them. The aborting of young fruits will occur as a self-pruning process.

**Harvesting** Pick summer squashes when they are young and tender. The seeds should be undeveloped and the rind soft. Pick continually for a steady supply of young fruit. Zucchini, straightneck, and crookneck types are usually picked at 1½ to 2 inches in diameter, and scallops at 3 to 4 inches across.

Winter squashes must be thoroughly mature to have good quality. When picked immature they are watery and poor in flavor. After some cold weather increases the sugar content, the flavor usually improves. Learn to judge varieties by color. Most green varieties have some brown or bronze coloring when they ripen in the fall.

Pumpkins are harvested in the fall when the rind becomes hard and the foliage starts to die. As the pumpkins mature, raise them off the ground on a board to prevent the bottom from rotting.

See Gourds on page 127 for further information on their harvest and use.

### Summer Squash Varieties

Summer squashes are generally divided into four types: crookneck, with tapered bodies and curved necks; scallop, which is round to plate shaped with scalloped edges; straightneck, long, straight, and tapering; and zucchini, straight and cylindrical.

#### Crookneck

'Early Golden Summer', bumpy bright yellow rind, 53 days.

#### Scallop

'Early White Bush', also called 'White Patty Pan', creamy white flesh and pale green rind, 60 days.

'Patty Green Tint', a rich-flavored hybrid with light green rind, 52 days.

'Peter Pan', light green fruit, an All-America Selection hybrid, 50 days.

'Scallopini', dark green zucchini color, scallop shape, productive, All-America Selection, 50 days.

'Sunburst', All-America Selection hybrid, bright golden yellow, lightly scalloped, marked in green at the edges, 53 days.

#### Straightneck

'Butterstick', hybrid with golden fruits with creamy white flesh, 50 days.

'Early Prolific Straightneck', big, bushy plant, cylindrical light yellow fruit, an All-America Selection, 52 days.

'Goldbar', golden yellow, 5 to 6 inches long, 53 days.

#### Zucchini

'Ambassador', hybrid with smooth dark green fruit, 7 to 8 inches long, 48 days.

'Aristocrat', dark green, smooth, cylindrical, an All-America Selection, compact plant, 50 days.

'Burpee Golden', glossy bright gold fruit, 54 days.

'Burpee Hybrid', shiny medium-green fruit, 50 days.

'Chefini', early, medium green, solid, an All-America Selection, 48 days.

'Gold Rush', deep gold color, smooth rind, good flavor, small plants, an All-America Selection, 52 days.

'Greyzini', light green with gray mottling, high yielding, All-America Selection, 47 days.

'Richgreen', hybrid with shiny dark green rind, very vigorous and high yielding, matures in 50 days.

Two varieties defy the above categories because of

'Gold Nugget' squash

Spaghetti squash

the unique shape of their fruit: 'Gourmet Globe', a hybrid that bears round dark green fruit with lighter markings, 4 to 6 inches across, 45 days; and 'Sun Drops', an All-America Selection, 3- to 4-inch oval yellow fruit on a compact plant, 55 days.

### Winter Squash Varieties

Winter squashes are divided into broad classifications: acorn, with primarily round, deeply ribbed dark green fruit; butternut, shaped like a large pear with a long, thick neck; hubbard, large, round to oval, ribbed, bumpy fruit; and turban, with flat, rounded fruit, some of which have a distinct center that resembles the headgear of the same name.

### Acorn

'Cream of the Crop', All-America Selection hybrid, bushy plant, beautiful fruits have cream-colored rind with golden flesh, 85 days.

'Jersey Golden', gold rind and light yellow to orange flesh, small plant, can be harvested and used like summer squash, 60 to 80 days.

'Table Ace', semibush hybrid, black rind, orange flesh, 70 days.

'Table King', glossy dark green fruit, golden flesh, a good keeper, All-America Selection, 75 days.

'Table Queen', semibush plant, dark green rind, orange flesh, nutty flavor, 90 days.

### Butternut

'Butterbush', reddish orange rind and flesh, space-saving bush type, 75 days.

'Early Butternut', hybrid, semibush, productive, tan rind, orange flesh, All-America Selection, 80 days.

'Waltham', All-America Selection, dark orange to tan rind, solid, dry flesh, 85 days.

### Hubbard

'Blue Hubbard', blue-gray rind, yellow-orange flesh, up to 15 pounds, All-America Selection, 120 days.

### Spaghetti Squash

A unique winter squash, 'Vegetable Spaghetti' has stringlike flesh that can be fluffed out of the rind after boiling and served as an excellent low-calorie substitute for spaghetti. The medium-sized oblong fruit takes about 100 days to mature.

### Turban

'Buttercup', sweet orange flesh, strong flavor, dark green–striped rind, 105 days.

'Sweet Mama', for small gardens; large fruits are dark gray-green, short vines are pinched off at 4 feet, deep gold, sweet flesh, cooks and stores well, All-America Selection, 85 days.

'Turk's Turban', fruit striped with bright orange or red, often listed as a gourd, 105 days.

### Pumpkin Varieties

'Autumn Gold', All-America Selection hybrid, bright yellow, 90 days.

'Big Max', wins prizes for size, not eating, often over 100 pounds, 120 days.

'Big Moon', 1 or 2 huge pumpkins per vine, each may weigh as much as 200 pounds, 120 days.

'Cinderella', bush, medium-sized jack-o'-lantern, 95 days.

'Connecticut Field' (or 'Big Tom'), large jack-o'-lantern, 120 days.

'Green Striped Cushaw', unusual, pear-shaped, creamy white rind with pale yellow flesh, the edible part is the neck, 90 days.

'Jack O' Lantern', medium sized, 110 days.

'Small Sugar Pie' (or 'New England Pie'), small, globe-shaped fruit, 100 days.

'Spirit Hybrid', medium-sized pumpkins on vines more compact than most, symmetrical fruits perfect for Halloween carving, All-America Selection, 100 days.

'Triple Treat', uniformly shaped for carving, good flesh for eating, produces hull-less seeds for eating, 100 days.

Pumpkin varieties developed specifically for seeds form no hulls around the seeds. In Mexico and Central America, pumpkin seeds are sold by street vendors, much as peanuts are in the United States. Two recommended varieties are 'Lady Godiva', with 8-inch fruits, 110 days; and 'Hungarian Mammoth', 115 days.

**How to use** Summer squash is good either cooked or raw. Slice it to cook in stir-fried dishes, quiches, and omelets.

French pumpkin

Gourd

Crooknecks look festive halved, steamed, buttered, and dusted with nutmeg. The pretty scallop squash takes well to stuffing. For barbecues, arrange summer squash slices on aluminum foil, drizzle with olive oil, and add salt and pepper, garlic powder, and oregano. Grill 5 inches above hot coals for about 20 minutes.

Zucchini, the most prolific of the summer squashes, has probably inspired more recipes than any single vegetable. The zucchini glut that strikes many gardens in midsummer can be handled in many delightful ways. Try sweet zucchini bread, pancakes, or cake. Split large ones and fill with ground beef, curried lamb, rice, seasoned bread crumbs, or a stuffing of your choice, and bake, topping with herbs and grated cheese during the last few minutes in the oven. Use slices of zucchini as dippers on a relish tray, or shred zucchini and mix with sour cream dressing for a colorful slaw.

With winter squash, the first step is cutting it down to serving size. Large squash may require a heavy knife or handsaw for the initial cut. Make sure to remove all seeds and stringy portions.

Winter squash can be steamed, baked, or broiled. To speed baking, cut squash in half and arrange halves on a pan cut side down. Bake until nearly tender, then turn right side up, add butter and seasonings, and bake another 15 to 20 minutes. Eat straight from the rind or scoop out the flesh and dress it up with cream, brown sugar, and butter. Winter squash takes well to sweet spices and seasonings and many garnishes.

Pumpkin can be prepared by steaming or baking, as with winter squash, then seasoned with butter, salt and pepper, and brown sugar or molasses. Try pumpkin bread, pie, or cake, glazing with corn syrup seasoned with lemon zest and ground ginger. For a treat on a nippy fall day, bake pumpkin cookies filled with raisins.

### Gourds

Gourds are a true gardening curiosity. Unusual in shape, color, and markings, gourds defy predictable results because they can cross with one another. The chart on the next page explains the relationship between different members of the gourd family and those plants that are specifically known as gourds.

If you look down the column headed Genus in the chart, you will come upon *Lagenaria* and its one species, *siceraria*. These are the larger gourds, often listed in seed catalogs as separate varieties, such as bottle, dipper, and 'Hercules Club'. In many regions of the world, before the invention of pottery, such gourds made up the total of domestic utensils, being fashioned into bottles, bowls, ladles, spoons, churns, and many other containers. They are still used to make musical instruments, pipes, and floats for fishing nets.

Seed growers have maintained the individual shapes and colorings of these ornamental gourds over the years by isolating the growing area for each. Different varieties will cross readily if grown together, and with time, present forms would change and new ones appear.

All gourds are fast growers if they have their quota of heat, especially at night. Delay planting until the soil is warm. In short-season areas start seed indoors in pots 3 to 4 weeks before the average last frost. Set out transplants or thin seedlings to 2 feet apart. Soil, water, and fertilizer needs are the same as for their relatives the squashes, melons, and cucumbers.

**Drying** Most gourds can be dried. Pick them when the stems turn brownish. Punch the end close to the stem with a long needle to allow air inside, then hang for several months in a well-ventilated place. The seeds will rattle when the gourds are fully dry. To make containers, cut them with a sharp saw and scrape out the insides; clean the rind with a pot scrubber and cover inside and out with several coats of shellac.

### Luffa

The luffa (*Luffa cylindrica*)—also known as the vegetable sponge, the dishrag gourd, and Chinese okra—grows rapidly to 10 to 15 feet, making cylindrical fruits 1 to 2 feet long.

## The Cultivated Members of the Gourd Family

Family: *Cucurbitaceae*

| Genus | Species | Variety or Common Name |
| --- | --- | --- |
| *Cucurbita* | *pepo** | 'Jack O' Lantern' pumpkin; 'Connecticut Field' pumpkin; acorn squash, all varieties; spaghetti squash; summer squash, all varieties; small, hard-shelled gourds; edible gourd; vegetable gourd |
| | *moschata** | Butternut squash, all varieties; pumpkin, some varieties |
| | *maxima** | Turban squash, all varieties; hubbard squash, all varieties; 'Big Max' pumpkin; 'Big Moon' pumpkin |
| | *ficifolia* | Malabar gourd |
| | *mixta** | 'Green Striped Cushaw' pumpkin |
| *Lagenaria* | *siceraria* | Bottle or white-flowered gourd; cucuzzi |
| *Luffa* | *cylindrica* | Dishrag gourd |
| *Momordica* | *balsamina* | Balsam-apple |
| | *charantia* | Balsam-pear |
| *Sechium* | *edule* | Chayote |
| *Benincasa* | *hispida* | Chinese preserving melon or white gourd of India |
| *Cucumis* | *melo* | Netted muskmelon or cantaloupe; honeydew & casaba muskmelons; crenshaw melon; oriental pickling melon; dudaim melon; snake or serpent melon; mango melon |
| | *sativus* | Cucumber, all varieties |
| | *anguria* | West Indian gherkin |
| *Citrullus* | *vulgaris* | Watermelon, all varieties; citron |
| *Trichosanthes* | *anguina* | Snake or serpent gourd |

* See text about crossing between these species.

Balsam-apple gourds

Luffa

'Turk's Turban' squash

Luffas are thought to have originated in tropical Asia. They reached China about A.D. 600, and are now cultivated throughout the tropics. Although they are tropical plants, the best luffas are grown in Japan.

To reach the spongy, fibrous interior, the ripe gourds are immersed in a tank of running water until the outer wall disintegrates. They are then bleached and dried in the sun. The luffa is grown commercially for use as sponges and in the manufacture of many products—filters in marine and diesel engines, bath mats, table mats, sandals, and even gloves.

In India the young, tender fruits are eaten raw like cucumbers or cooked as a vegetable. In Hawaii and China the small pods are used to replace edible-pod peas in chop suey.

**How to use** Luffas must grow to maturity to be used for sponges, but for eating they should be picked when no more than 4 to 5 inches long.

Slice them raw into salads or butter-steam as you would summer squash. They adapt to most zucchini recipes and combine especially well with tomatoes. Use the leaves in salads, or cook as greens. The flowers can be dipped in batter and deep-fried.

To make homegrown scrubbers, soak luffas in water for several days until the skin falls off. Then dry them in the sun.

### Snake Gourd

Although the snake gourd, or serpent cucumber, grows up to 6 feet in length, it's best for eating when harvested at a fraction of that size. Snake gourds taste like cucumbers when eaten raw. When cooked they resemble zucchini in flavor and go well in almost any recipe for summer squash. They are a prized delicacy in much of Asia.

### Edible Gourds

Closely related to both squashes and pumpkins, the vegetable gourd is grouped in the *C. pepo* species. A vigorously growing vine, it has attractive foliage and fruits, which are displayed to advantage when trained on a trellis. The fruits are shaped like miniature pumpkins, 3 to 5 inches across, and weigh about ½ pound. When mature the fruit is striped a creamy white with dark green mottling.

Vegetable gourds taste like sweet winter squash. They can be stuffed like bell peppers with meat or rice, then baked; or boiled and mashed like winter squash.

There are also edible gourds that are varieties of *Lagenaria siceraria*. For the best eating, these edible gourds should be harvested young, while the fuzz is still on them. They have a rich, full flavor and can be cooked just about any way you would prepare summer squash or eggplant. *Lagenarias* are very popular in Italy, so try them Italian style, baked with fresh tomatoes and sprinkled with basil and olive oil.

### Crossing Gourds and Squashes

Except for perhaps the mustard family, the gourd family has, among the vegetables, the greatest diversity in its edible forms, and certainly the widest variation in color and form of fruit. Cross two varieties of summer squash, such as zucchini or yellow crookneck with white bush scallop, and the second generation will display an unbelievable array of color, shape, texture, and size of fruit. In a population numbering in the hundreds, no two will be alike.

The pumpkin and squash seeds you buy are from varieties grown in areas free from the pollen of any other variety. However, nature has a way of sneaking in a cross or two. These will show up as occasional strange plants in the garden. Seeds saved from these will produce many different forms the next year.

The possibilities for crossing can be summarized as follows. Any two varieties of the same species will cross freely. For example, varieties of hubbard and turban squash, which are both *Cucurbita maxima,* will cross, as will varieties of muskmelon and Crenshaw melon, which are *Cucumis melo.*

Crossing between species does occur in the genus

Bottle gourds

'Pygmy Dwarf' sunflowers

*Cucurbita*. Both *C. pepo* and *C. maxima* will cross with *C. moschata*, but *C. pepo* and *C. maxima* will not cross with each other. An additional cross, *C. pepo* with *C. mixta*, will occur. All of this means, for example, that acorn and butternut squashes will cross, but acorn squash will not cross with hubbard squash.

Other crosses between species, such as muskmelon with cucumber, do not occur. Nor do any crosses occur between one genus and another.

## SUGAR PEAS OR SNOW PEAS

See Peas.

## SUNFLOWERS *Helianthus annuus*

Stand next to this old-fashioned garden annual when it is full grown and you'll feel as if you've stepped into a land of giants. Growing as tall as 12 feet, with flower heads as large as 18 inches across, sunflowers stand like sentries, topped with huge sunbursts of golden yellow. Plant them at the back of the flower border or in the vegetable garden. Each plant will supply a bounty of seeds for birdseed or good eating.

Some gardeners grow sunflowers as windbreaks, or plant the seeds 2 weeks before pole beans to let the stalks support the climbing vines.

Sow the seeds ½ inch deep when the soil is warm and after frost danger has passed. Thin, depending upon the size of the variety, to 12 to 48 inches apart. Fertilize little if any, and water sparingly. If you have a problem with birds snatching the seeds, cover the flower heads with cheesecloth and tie it securely at the back of the flowers.

**Varieties** There are many sunflower varieties available, ranging from a few inches in height to more than 12 feet. Some form a single large head and others many smaller heads. The plump, meaty seeds of 'Mammoth', 6 to 12 feet tall, can be harvested in 80 days. 'Sunspot' has 12-inch flowers on 24-inch plants, 80 days.

**How to use** Cut flower heads when the seeds become hard, and hang them to dry in a location with good air circulation. Tie a cloth bag around each head to catch any seeds that might drop during drying. Seeds may be eaten fresh or roasted. To roast, mix thoroughly 2 cups unwashed dried seed, ½ teaspoon Worcestershire sauce, 1½ tablespoons melted butter, and 1 teaspoon salt. Place in a shallow baking pan and roast for 1 hour at 250° F. To ensure even browning, shake the pan several times. Place seeds in a plastic bag and store in the refrigerator. Squash and pumpkin seed can be roasted in the same way.

## SWEET POTATOES *Ipomoea batatas*

This member of the morning glory family was taken from Central and South America to Spain by Columbus. After the conquest of Mexico, the Spaniards took sweet potatoes to the Philippines, and the Chinese adapted them there. Records show they were being cultivated in what is now Virginia in 1648.

No vegetable commonly grown in the United States will withstand more summer heat than the sweet potato; indeed, very few require as much heat. This tropical plant does not thrive in cool weather. A light frost will kill the leaves and soil temperatures below 50° F will damage the tubers.

Commercial crops are feasible where mean daily temperatures (average of day and night) are above 70° F for at least 3 months. Louisiana, North Carolina, and Georgia all produce significant commercial crops. In the West, only the hottest summer areas of Arizona and California support commercial culture.

Although the names sweet potato and yam are often used interchangeably, they are different plants, and differ in both growth habit and culture. Yams belong to the genus *Dioscorea* and are rarely grown outside of the tropics. They are not as nutritious as sweet potatoes, being little more than starch. Shredded and cooked, they become mucilaginous, or gluey.

Fertilizing sweet potatoes is tricky. Given too much nitrogen, they develop more vines than tubers. However,

Sweet potatoes

Sweet potato plant in bloom

they are not a poor-soil crop, and a low-nitrogen fertilizer such as 5-10-10 worked into the soil at the rate of 4 pounds per 100 feet of row will improve the yield. Prepare the soil 2 weeks before planting. In some areas potash will increase yields and make the tubers shorter and chunkier; low potash likely means long, stringy potatoes.

**How to plant** Sweet potatoes are started from slips. Supermarket tubers may sprout and produce the slips, but like Irish potatoes they are usually treated to prevent sprouting. Disease-free slips from a garden center or mail-order seed company are best, and you'll be sure of the variety.

Plant promptly after the soil is warm, since the length of the growing season is often the limiting factor. Where the season is long enough, plant 2 weeks after the last frost to ensure thoroughly warm soil.

If drainage is poor or plants are overwatered, the tubers may be elongated and less blocky. To ensure good drainage, sweet potatoes normally are planted in 6- to 12-inch ridges, located 3 to 4½ feet apart. Where soil is fast draining and sandy, ridge planting is of little advantage.

Space the plants 9 to 12 inches apart. Rich soil may allow closer spacing. If spacing is increased much beyond 24 inches, more oversized tubers will result but at the expense of both total yield and quality.

Being deep-rooted plants often grown in moisture-poor sandy soils, sweet potatoes are usually able to find enough water to survive; but those planted for harvest should never have to struggle for water. As a general guide, they will use about 18 inches of water per season.

After your first sweet potato crop, save some of the healthiest tubers to produce your own slips. Cut or split the crown and underground stems of each plant and examine for dark strands or general darkening of internal tissue, a symptom of stem-rot infection. Do not use tubers from stems whose internal tissue shows any discoloration, even though the sweet potatoes appear fine. Also examine for surface cracks and black eyes, which indicate nematode infestation.

Three potatoes should produce about 24 slips, enough for a 25-foot row and about 60 pounds of tubers. Plant the chosen tubers close together in sand, vermiculite, or perlite hotbeds 5 to 10 weeks before your outdoor planting date. When sprouts reach 9 to 12 inches, cut them off 2 inches above the soil, and set the slips into containers with a high-quality, preferably sterile rooting soil. Keep the soil warm (80° to 90° F) for fast rooting. Sufficient roots will usually form in 10 to 14 days.

**Harvesting** Harvest sweet potatoes when slightly immature as soon as the tubers are large enough. Otherwise wait until the vines begin to yellow. Try to avoid bruising them when digging, since this invites decay. If the leaves are killed by frost, harvest immediately.

Sweet potatoes improve during storage, because part of their starch content turns to sugar. For storage, they need to be cured. Let the roots lie exposed for 2 to 3 hours to dry thoroughly, then move them to a humid (85 percent relative humidity) and warm (85° F) storage area. After 2 weeks lower the temperature to 55° F, in which they will keep between 8 and 24 weeks.

**Beginners' mistakes**
Planting too late to take advantage of the full growing season is the most common error. But don't plant too early either. If soil is not at least 50° F, plants will languish, if not perish.

**Limited space** If you lack the garden space for sweet potatoes but would still like to grow some, try them in a box at least 12 inches deep and 15 inches wide. Use a light, porous soil mix and place a 4-foot stake in the center to support the vine. Or grow them as a lush, vining houseplant in a bowl or jar.

**Varieties** Sweet potatoes in the United States mean two basic types of vegetable: the pale yellow, slightly sweet, fairly dry northern version; and the dark orange, moist, distinctly sweet southern type. The southern "sweets" are those commonly and erroneously called yams. Dry-fleshed varieties include 'Jersey Orange', 'Nugget', and

Sweet potato sprouts started in water

Tomatillo

'Nemagold'. Moist-fleshed varieties include 'Centennial', 95 days, good for short growing seasons; 'Porto Rico', 125 days, a short-vined type good for small gardens; and 'Gold Rush', 125 days, resistant to wilt.

**How to use** Most cookbooks use the two types of sweet potatoes interchangeably, but common sense dictates using less sweetener with moist-fleshed yams and using more butter or cream on the drier sweet potatoes. Dieters should be aware that yams are much higher in calories.

Both vegetables are delicious baked in the skin and served with butter. Glaze them with brown sugar or maple syrup, or with orange juice (and a little grated zest) or crushed pineapple. Try baking sliced boiled sweet potatoes with slices of apples, garnished with raisins and chopped nuts.

Sweet potatoes and yams complement pork chops, lamb, turkey, and ham, and also find their way into a number of baked treats, from biscuits to sweet potato pie and cake.

## TOMATILLO
## *Physalis ixocarpa*

The tomatillo is a perennial, often grown as an annual, reaching 3 to 4 feet high. The leaves are long, oval, and deeply notched. The fruit is smooth, sticky, either green or purplish 1 to 2 inches in diameter, and entirely enclosed in a thin husk that looks like a Chinese lantern.

A close relative of the husk tomato, the tomatillo has a tart flavor—to some, similar to green apples. Tomatillos and green tomatoes are never interchangeable in recipes. The flesh is of a different texture, the former seedy but solid, without the juicy cavities of the tomato.

Grow tomatillos the same way as tomatoes. Seed sown in peat pots will germinate in about 5 days and be ready to transplant in 2 to 3 weeks. Tomatillos are ready to harvest in about 100 days.

Harvest tomatillos according to the intended use. For highest quality fruit, harvest when the husks change color from green to tan; otherwise, they lose their tartness and become soft. Left on the vine, they become yellow and mild. They can be stored for months. Some gardeners store them on the vine; others spread out the picked fruit, still in their husks, in a cool place with good air circulation. (Packed into airtight plastic bags, the berries spoil rapidly.)

**How to use** Tomatillos are used in many Mexican recipes. Raw or cooked, they give sauces a rich, distinctive flavor. Try them fresh in salads, tacos, and sandwiches. Tomatillos are probably best known for the contribution they make to Mexican *salsa verde,* or green taco sauce. Add chiles for hotness.

## TOMATOES
## *Lycopersicon lycopersicum*

None of the vegetables from the New World took as long to be appreciated in the Old World as the tomato. Used for centuries in Mexico and Central and South America, the tomato is recorded as being cultivated in France, Spain, and Italy in 1544; but a century later it was being grown in England only as a curiosity.

The first tomato seeds to reach Europe were of the yellow variety. They became the *pomi d'oro* (apples of gold) of Italy, and a few years later, the *pommes d'amours* (apples of love) of France.

In pioneer America only a few would venture to eat the fruit, which was thought poisonous by many. New Englanders in Salem in 1802 wouldn't even taste tomatoes; they were finally recognized as a useful vegetable by 1835.

Tomatoes are warm-season plants and should be planted at least 1 week after the average last frost. Temperature is a most important factor; tomatoes are particularly sensitive to low nighttime temperatures. In early spring when daytime temperatures are warm but nights fall below 55° F, many varieties will not set fruit. In summer you can expect blossom drop when days are above 90° F and nights above 76° F.

In the variety chart on page 136, you will note varieties suitable for northern and southern climates.

The soil for tomatoes should be well drained and

Tomato plant grown in wire cage

Tomato plants behind wind protector

have a good supply of nutrients, especially phosphorus. To prepare the soil, use plenty of organic matter and add 3 to 4 pounds of 5-10-10 fertilizer per 100 square feet. Water it in and allow 2 weeks before planting.

Most gardeners start tomatoes with transplants, which are usually available at garden centers at the earliest possible planting time. To start your own transplants, sow seed ½ inch deep in peat pots or other plantable containers 5 to 7 weeks before the outdoor planting date. See page 54 for information on starting seeds. The last 10 days before planting outdoors, gradually expose the seedlings to more sunlight and outdoor temperatures.

In the garden, tomato seed is hard to germinate, and limited success is found only in long growing seasons.

Purchased transplants should be stocky, not leggy, and should have 4 to 6 young and succulent true leaves. Avoid plants already in bloom or with fruit, especially if they are growing in very small containers.

Set transplants deep, the first true leaves just above soil level. Plant leggy plants with the rootball horizontal. Roots will form along the buried stem and make better subsequent growth. If cold or wind are threats, use hot caps or other protection.

Determinate (bush type) and ISI (compact bush type) tomatoes should be spaced 24 inches apart. Staked indeterminate (vine type) tomatoes should be spaced 18 inches apart, whereas those that grow in cages need 2½ to 3 feet between plants. Indeterminate tomatoes grown on the ground need 4 feet between plants.

The first fertilizer application will take care of the plant until it sets fruit, at which point it should be fed again. Then feed once a month while the fruits are developing. Stop when they near mature size.

Tomatoes require uniform moisture after the fruit has set; alternate wet and dry spells can bring on stunting and blossom-end rot. In the early stages you can stretch watering intervals to put the plant under a little stress; it's a good way to bring on tomato production if you're careful not to overdo it. When harvest time is near, cut back slightly on watering to get less watery fruit and better flavor.

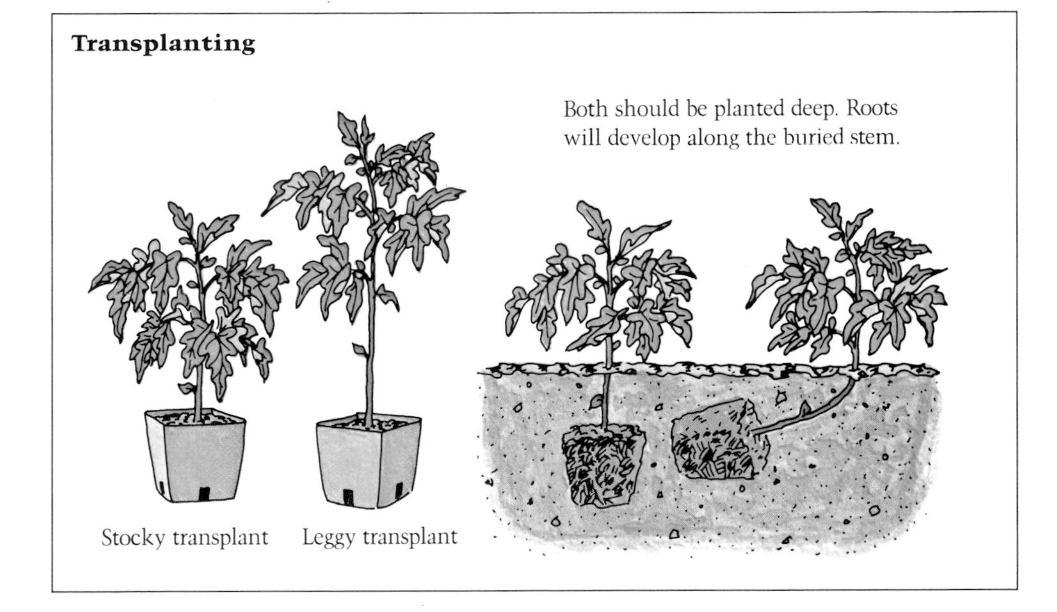

'Beefmaster' tomatoes

'Big Early' tomatoes

**Tomatoes in containers**
Any tomato that can be grown in a vegetable garden can also be grown in a container. Gardeners with unfavorable soils find especially that growing in containers with a disease-free planting mix is worth the extra care in watering and feeding.

Grow dwarf plants in 8-inch pots. Several varieties can be grown in hanging baskets. Giant-sized plants will thrive even in 2-gallon containers of planting mix if you compensate for the limited root space with extra water and fertilizer. Use any container you like, but be sure to provide drainage.

**Training** No vegetable responds better to training than the tomato, and few plants are trained as easily. They can be grown on upright stakes and trellises, in wire cages, or on horizontal trellises or ladderlike frames set a foot above ground level. All will keep fruits from contact with the soil and reduce damage from slugs, cracking, sunscald, and decay. In wet fall climates the yield of usable tomatoes can nearly be doubled by training the fruit on supports.

Although indeterminate tomatoes can be grown on the ground and will produce more fruit than if grown on a support, the fruit will be smaller and will take longer to ripen.

The low-growing, bushy varieties are difficult to stake but may be held up by horizontal frames of different kinds. The plants may be allowed to sprawl but should then be protected by mulch, either organic or plastic.

Tall-growing varieties may be grown on 6-foot stakes set a foot deep into the soil. They can also be pruned to one stem and grown on strings strung on a wooden frame.

**Pruning** The market gardener aiming for the early market should sacrifice quantity for quality and prune to a single stem, removing all suckers. This shortens the growing season and ripens fruit uniformly.

The home gardener can modify this system and harvest more fruit over a longer period by allowing one sucker to grow from near the base to form a two-stemmed plant, and later removing the rest of the suckers on both stems.

One disadvantage of heavy pruning is a lack of foliage to shade the fruit and protect it against sunscald. Pruned plants are also more prone to develop blossom-end rot.

A way to achieve early fruit production and later sun protection is to take out all suckers on the lower 18 inches of the stem, then let the plant bush out with the branches tied to a support.

**Protection** Early planting calls for protection against low temperatures. If you use a wire cage, a cover of polyethylene film in the early stages of growth will raise the temperature inside the cage. A row of 4 or 5 plants can be covered with a 2-foot-high polyethylene film row tent and later with a tent as high as the mature plants.

**Tomato Problems**

Here is a quick summary of some of the problems you might run into when growing tomatoes. These aren't diseases, but just problems caused by climate or care, and can usually be avoided with the right care procedures.

**All vine and no fruit** If the plant doesn't produce flowers, the problem is too much nitrogen and water in the early growth stages. Too much nitrogen stimulates vigorous vine growth but delays maturity. Dry the plants up a bit to try to induce flowering. If flowers are being formed, but they are dropping off before they set fruit, the problem is blossom drop.

**Blossom drop** For a tomato grower this can create great anxiety. The blossoms are out, but the big question is, will they drop or set fruit? To find out takes about 50 hours—the minimum time required for pollen to germinate and the tube to grow down the pistil to the ovary. At nighttime temperatures below 55° F germination and tube growth are so slow that blossoms drop off before they can be fertilized. As a rule, most early-maturing varieties set fruit at lower temperatures than the main-season kinds.

Fruit set can also be hampered by rain and prolonged humid conditions. Growers in cool, humid situations have found that fruit set can be increased by shaking the plant, or vibrating it with a battery-powered toothbrush, to release

'Pink Girl' tomatoes

Cherry tomatoes

pollen for pollination. With stake-trained plants, hitting the top of the stakes will accomplish the same end.

**Blossom-end rot** Symptoms of this disease appear as a leathery scar or rot on the blossom end of fruits. It can occur at any stage of development and is usually caused by sudden changes in soil moisture, most serious when fast-growing plants are hit by a hot, dry spell. Lack of calcium is another cause.

Mulching with black plastic or an organic material, which reduces fluctuations in soil moisture and temperature, and avoiding planting in poorly drained soil will help prevent blossom-end rot. Also, staked and heavily pruned tomatoes seem more susceptible to the disease than unpruned plants.

**Curled leaves** Wilt during a hot spell at midday is normal. Plants in containers show top-growth wilt and drooping when they need water. Once watered, they recover rapidly.

Some kinds of leaf curl are normal. It's more pronounced in some varieties, and you can expect it during hot, dry spells and during and after a long wet period. Heavy pruning also seems to encourage leaf curl.

**Poor fruit color** In hot-summer areas, high temperatures can prevent the normal development of fruit color. The plant's red pigment does not form in temperatures above 86° F. Uneven coloring is common if fruits mature in high temperatures.

Both high temperatures and high light intensities will stop the color from forming in fruit exposed to the direct sun, and fruits may sunscald. Where high temperatures are the rule, choose varieties with a dense foliage cover.

### Beginners' Mistakes

- Failure to fit the variety to the climate.
- Failure to choose disease-resistant varieties.
- Planting too early.
- Trying to grow tomatoes in a shady location. They require at least 6 hours a day of direct sunlight.

**Varieties** The rule of thumb in choosing tomato varieties to fit your garden is this: The shorter the growing season, the more you should limit your choices to the early and early midseason varieties. Finding them should be no problem, as you can see by looking at the variety chart on pages 136 and 137.

In checking the varieties listed in the chart, note especially those with resistance to soil-borne diseases, indicated by V for verticillium wilt, F for fusarium wilt, N for nematodes, T for tobacco mosaic virus, and A for alternaria. Scientists have isolated two types of fusarium disease. A variety marked F is resistant to one type; those marked FF are resistant to both types.

It may be that your soil is not infested with any of these diseases and you can successfully grow any variety. But if you've had any trouble with tomatoes in the past, favor the resistant varieties.

The number of days shown on the chart refers to the time from setting out transplants to the first fruits. It is an average figure and intended only as a general guide. The actual number of days a plant needs to reach maturity in your garden depends on the climate.

The growth habit of each variety is indicated by the words *determinate* or *indeterminate*. The determinate are the bush kinds, generally growing to 3 feet or less. All the fruit ripens at the same time. Determinate tomatoes are usually grown by gardeners who want a quantity of fruit all at one time to can or otherwise process. The indeterminate are tall-growing, vining types and are trained with stakes, a trellis, or a wire cage. A healthy plant will produce fruit until it is killed by frost. In recent years a new type of tomato called intermediate short internode (ISI) has been developed. It flowers and fruits indefinitely, like the indeterminate tomato, but grows as a compact bush, like the determinate kinds.

Tomatoes come in a wide range of fruit and plant sizes. Most tomatoes are red, but there are pink and yellow selections as well.

**How to use** To many tomato lovers, the supreme tomato is the one plucked from the vine on a midsummer afternoon and eaten right in the garden. For other gardeners, growing

## Tomato Varieties

| Variety | Hybrid | Days to Maturity | Growth Habit* | Fruit Size | Disease Resistance* | Comment |
|---|---|---|---|---|---|---|
| **Large Fruited** | | | | | | |
| 'Ace 55' | | 80 | Det | Large | VFA | Adds disease resistance but reduces fruit smoothness compared to standard 'Ace'. |
| 'Ace-Hy' | • | 76 | Det | Large | VFN | Smooth fruit; vigorous, productive. |
| 'Basket King' | • | 55 | Det | Small | | Cascading branches, excellent for hanging baskets. |
| 'Beefmaster' | • | 80 | Ind | Very large | VFNA | Beefsteak type with triple disease resistance, vigorous. |
| 'Beefsteak' | | 96 | Ind | Large | | Ribbed, irregular, and rough fruit; vigorous vine with coarse foliage. |
| 'Better Boy' | • | 75 | Ind | Large | VFNA | Widely adapted; 8- to 16-ounce uniform fruit, heavy cropper, rugged. |
| 'Better Bush' | • | 72 | ISI | Large | VFN | Full-sized fruits on a compact plant. |
| 'Better Girl' | • | 62 | Ind | Medium | VFN | Meaty, crack-resistant fruits, heavy yielding. |
| 'Big Boy' | • | 78 | Ind | Very large | | Smooth, firm, thick walled, productive. |
| 'Big Early' | • | 62 | Ind | Large | | Early, solid fruit, bright red. |
| 'Big Girl' | • | 78 | Ind | Large | VF | Same as 'Big Boy' but with VF resistance. |
| 'Big Pick' | • | 70 | Ind | Very large | VFFNTA | Round, smooth fruit. |
| 'Bigset' | • | 70 | Det | Large | VFFNA | Sets well in both high and low temperatures, vigorous, resists cracking and blossom-end rot. |
| 'Bonus' | • | 75 | Det | Med.-lg. | VFN | Smooth, firm, catface-resistant fruits. Sets well in high temperatures. |
| 'Bragger' | • | 75 | Ind | Very large | | Strong grower, red and meaty fruit, very crack resistant. |
| 'Burpee's VF' | • | 72 | Ind | Med.-lg. | VF | Widely adapted; vigorous vine and meaty, crack-resistant fruit. |
| 'Cal-Ace' | | 80 | Det | Large | VFA | Sets heavily, fruit smooth. |
| 'Carnival' | • | 70 | Det | Large | VFFNTA | Thick-walled, uniform fruit. |
| 'Celebrity' | • | 72 | Det | Large | VFFNTA | Vigorous grower, All-America Selection. |
| 'Champion' | • | 62 | Ind | Large | VFNT | High yielding, large and early. |
| 'Early Cascade' | • | 68 | Ind | Medium | VA | Early, tall, cascading, and productive. |
| 'Early Girl Improved' | • | 52 | Ind | Small | VFF | Earlier, more disease resistance than 'Early Girl'. |
| 'Early Pick' | • | 62 | Ind | Medium | VF | Similar to 'Big Early', better fruit set, disease tolerance. |
| 'Fantastic' | • | 72 | Ind | Med.-lg. | A | Productive, widely adapted and recommended. |
| 'Floramerica' | • | 70 | Det | Large | VFFA | All-America Selection, tolerance or resistance to 17 diseases or disorders. |
| 'Golden Boy' | • | 80 | Ind | Large | A | Mild flavor, bright yellow fruit. |
| 'Heinz 1350' | | 75 | Det | Med.-lg. | VF | Strong, compact vine, uniform 6-ounce fruit. |
| 'Homestead 24' | | 80 | Det | Large | FA | Sets under variety of conditions, including high temperatures. |
| 'Jet Star' | • | 72 | Ind | Med.-lg. | VF | Important variety in the second early season, sets well, good production. |
| 'Jubilee' | | 80 | Ind | Large | A | Golden-orange fruit, mild flavor. |
| 'Lemon Boy' | • | 72 | Ind | Medium | VFNA | Lemon-colored flesh, very flavorful. |
| 'Manalucie' | | 80 | Ind | Large | FA | Firm, meaty fruit; vine vigorous, upright; a hot climate favorite. |
| 'Marglobe Improved' | | 75 | Det | Medium | VF | Improvement over the longtime favorite 'Marglobe Select'. |
| 'Marion' | | 78 | Ind | Medium | FA | One of the best home-garden stake varieties, widely recommended, especially in hot climates. |
| 'Monte Carlo' | • | 75 | Ind | Large | VFNA | Tall, strong vine, productive over long season. |
| 'Mountain Pride' | • | 77 | Det | Med.-lg. | VFFA | Well adapted, strong vines. |
| 'New Yorker' | | 66 | Det | Medium | VA | Sturdy plants reliably produce heavy crops, good short-season variety. |
| 'Oxheart' | | 80 | Ind | Large | A | Heart-shaped fruits, pinkish, firm, meaty, and solid. |
| 'Patio' | • | 70 | Det | Sm.-Med. | FA | Dwarf plant, good for containers. |
| 'Pearson Improved' | | 78 | Det | Large | VF | Somewhat open vines with good disease resistance. |
| 'Pink Girl' | • | 76 | Ind | Large | VFTA | Pink skin, crack resistant. |
| 'Pink Ponderosa' | | 80 | Ind | Very large | | Old-timer, meaty and firm pink flesh; grow in cage. |
| 'Porter' | | 78 | Ind | Small | | Consistently produces over wide range of climate and soil conditions, good heat resistance, pink flesh. |

* See page 135 for explanation of abbreviations.

## Tomato Varieties (continued)

| Variety | Hybrid | Days to Maturity | Growth Habit* | Fruit Size | Disease Resistance* | Comment |
|---|---|---|---|---|---|---|
| 'Porter Improved' | | 65 | Ind | Medium | | Same vigor and reliability as 'Porter', with improved fruit color and size. |
| 'President' | • | 68 | Det | Large | VFFNTA | Productive plants, smooth fruit. |
| 'Quick Pick' | • | 52 | Ind | Medium | VFFNTA | Heavy yielding, for slicing or canning. |
| 'Ramapo' | • | 72 | Ind | Med.-Lg. | VFA | From Rutgers U., sets well under adverse conditions, strong and vigorous. |
| 'Rutger's Select' | | 75 | Det | Medium | VFA | Descendant of original 'Rutgers' and similar to 'Ramapo', widely available. |
| 'Spring Giant' | • | 65 | Det | Large | VFN | First All-America tomato, 1967; high yielding, concentrated harvest season. |
| 'Springset' | • | 65 | Det | Medium | VFA | Vigorous, open vine; susceptible to sunscald; good crack resistance. |
| 'Super Beefsteak' | | 80 | Ind | Very large | VFN | Smooth, flavorful, meaty fruit. |
| 'Super Bush' | • | 80 | Det | Medium | VFN | Compact, excellent for small gardens, no staking needed. |
| 'Super Fantastic' | • | 70 | Ind | Large | VFN | Smooth, solid fruits. |
| 'Super Sioux' | | 70 | Ind | Medium | | Noted for high-temperature fruit-setting ability; best in wire cage. |
| 'Supersonic' | • | 79 | Ind | Large | VF | Reliable main cropper in northern areas, crack resistant. |
| 'Supersteak' | • | 80 | Ind | Very large | VFN | Consistent texture, shape, and size; good for slicing, beefsteak type. |
| 'Terrific' | • | 73 | Det | Large | VFNA | Strong grower, produces over long season; good crack resistance. |
| 'The Juice' | • | 65 | Det | Large | | Excellent for canning or juice. |
| 'Vineripe' | • | 80 | Ind | Large | VFN | Heavy yields, vigorous grower. |
| 'Walter' | | 75 | Det | Medium | F | Vigorous and reliable, must be harvested ripe, good for hot climates. |
| 'Whopper' | • | 70 | Ind | Large | VFNT | Sweet, juicy, tasty, high yielding. |
| 'Wonder Boy' | • | 80 | Ind | Med.-lg. | VFN | Strong vine, medium foliage cover, heavy yield. |
| **Small Fruited** | | | | | | |
| 'Florida Basket' | | 70 | Det | ¾ in. | | Containers and hanging baskets. |
| 'Gardener's Delight' | | 65 | Ind | ¾ in. | | Prolific and crack resistant; one of the sweetest. |
| 'Pixie II' | • | 52 | Det | 1¾ in. | VFT | Early fruit on 14- to 18-inch plants; plant in 8-inch pot or hanging basket. |
| 'Red Cherry Large' | | 75 | Ind | | | Excellent yield of quality fruit. |
| 'Small Fry' | • | 65 | Det | 1 in. | VFNA | Cherry type in clusters on 30- to 40-inch vine; All-America Selection 1970; widely available. |
| 'Sugar Lump' | | 65 | Det | 1 in. | | Cherry-sized fruit on 30-inch vine for hanging basket or short trellis; unusually sweet. |
| 'Sweet 100' | • | 70 | Ind | 1 in. | | Very large, multiple-branched clusters of very sweet fruit; prune to one stem and stake or cage. |
| 'Tiny Tim' | | 60 | Det | ¾ in. | A | Small scarlet fruit on 15-inch plant; plant two in an 8-inch pot or hanging basket; widely available. |
| 'Toy Boy' | • | 58 | Det | 1½ in. | VFA | Plant three or four in a single 10-inch pot or hanging basket, indoors or out with plenty of light. |
| 'Tumblin' Tom' | • | 72 | Det | 1½ in. | | Early and heavy yield from 1- to 2-foot plants; good for hanging basket or window box. |
| 'Yellow Pear' | | 78 | Ind | 1 in. | A | Clear yellow fruit, mild flavor, medium to large plant. |
| 'Yellow Plum' | | 78 | Ind | 1 in. | A | Best yellow plum tomato for cooking. |
| **Paste Tomatoes** | | | | | | |
| 'Chico III' | | 75 | Det | Sm.-med. | F | Compact and slightly open growing, good for juice, sets fruit at high temperatures. |
| 'La Roma' | • | 62 | Det | Small | VFFA | More productive than 'Roma'. |
| 'Mama Mia' | • | 62 | Det | Small | VFF | Pear-shaped fruit, very productive. |
| 'Roma VF' | | 76 | Det | Small | VFA | Strong grower with dense foliage cover, large plant, pear-shaped fruit. |
| 'Royal Chico' | | 75 | Det | Small | VFNA | Growth habit improved compared to 'Roma', medium to large plant. |
| 'San Marzano Large Fruited' | | 80 | Ind | Small | A | Larger than San Marzano, pear-shaped, dry fruit, large plant. |

* See page 135 for explanation of abbreviations.

'Yellow Pear' tomatoes

'La Roma' tomatoes

the tomato is only a beginning: They're intrigued by the limitless possibilities of tomatoes in the hands of a good cook.

Ripe tomatoes should not be kept in the refrigerator for any length of time. You'll get top flavor if you store them in a cool place, as close to 60° F as possible. To peel, cover them with boiling water for 10 seconds, immerse in ice water until cool, then remove the skin with a serrated knife. Avoid peeling or cutting tomatoes until just ready to use.

Raw tomatoes can be sliced and served on an antipasto tray; marinated with sliced cucumbers in a vinaigrette dressing; combined with zucchini, hard-cooked egg, and chiles; or added to any tossed salad. Or stuff them with such fillings as tiny shrimp, potato salad, cold salmon, seasoned cottage cheese, or mashed deviled egg.

To ensure the best color, flavor, texture, and food value, cook tomatoes in a small amount of water until just tender; tomatoes create their own liquid as they simmer, and there's no point in diluting the nutrients. Cooked tomatoes are used in omelets, stews, casseroles, meat sauces, salad dressings, and even breads. You can create your own tomato juice and spaghetti sauce, use them for paste, or make fresh cream of tomato soup. Or enjoy gazpacho, a spicy cold soup of tomatoes, cucumbers, onions, and bell peppers, topped with a dob of sour cream.

Tomatoes are also excellent cooked as a single vegetable dish. Try them stewed, scalloped, stuffed, and baked (spinach makes a tasty filling), or sliced and glazed with wine and brown sugar. For barbecues, skewer tomato slices and grill them quickly over hot coals. Broiled tomatoes topped with butter and oregano make a mouthwatering accompaniment to a sizzling steak.

With the danger of approaching frost, many gardeners find themselves with an abundance of green tomatoes, which must be used quickly or not at all. Although green tomatoes will ripen indoors on a windowsill, channel some of them straight to the kitchen. You'll soon realize you've discovered a totally new vegetable.

Green tomatoes are delicious sliced, dipped in batter and bread crumbs, and fried in hot oil. They make tangy relishes, such as piccalilli, which can be enjoyed all winter, or they can be turned into mincemeat for use in cookies, cakes, and pies. Add them to orange marmalade for tartness or make simple green tomato jam. To many cooks, the best dish of all is green tomato pie, often compared to rhubarb or apple pie in flavor. Serve this distinctive dessert warm and topped with vanilla ice cream.

Putting up canned tomatoes or tomato preserves is one of the best ways to handle a sudden tomato glut. Canning is a safe, straightforward process if you find good canning instructions and follow them. Much has been written in the press about the risk of botulism from home-canned tomatoes, particularly from varieties low in acid content. The USDA has collected considerable data in an effort to ascertain the possible dangers. According to its report, "The seriousness of botulism should never be underestimated. However, we believe that the risk to home canners of tomatoes is very small. The home canner should not be overly concerned about hazards associated with the selection of specific tomato varieties. It is far more important to select tomatoes which are not overripe, to follow the recommendations of reliable canning guides explicitly, and destroy (without tasting) any home-canned product which appears abnormal in any way."

For insurance that the tomatoes you can are in the low pH range (or high in acidity), add ¼ teaspoon citric acid or 1 tablespoon lemon juice per pint of tomato product.

## TURNIPS AND RUTABAGAS

See Root Crops.

## YARD-LONG BEANS OR ASPARAGUS BEANS

See Beans.

## WATERMELON

See Melons.

## ZUCCHINI

See Squashes, Pumpkins, and Gourds.

## *Seed Sources*

It's generally best to purchase seeds and other garden material from a local nursery or garden center, if possible. However, if a local nursery doesn't carry the varieties you want, here are some mail-order nurseries that carry a wide variety of vegetable seeds. All have catalogs; some are free and some are available for a small fee.

**Agway, Inc.**
Box 4933
Syracuse, NY 13221

**Alberta Nurseries & Seeds Ltd.**
Box 20
Bowden, Alberta T0M 0K0
Canada

**Applewood Seed Company**
5380 Vivian Street
Arvada, CO 80002

**Archias Seed Store Corporation**
106 East Main
Sedalia, MO 65301

**Burgess Seed & Plant Company**
905 Four Seasons Road
Bloomington, IL 61701

**W. Atlee Burpee Company**
300 Park Avenue
Warminster, PA 18974

**D. V. Burrell Seed Growers Company**
Box 150
Rocky Ford, CO 81067

**Comstock, Ferre & Company**
263 Main Street
Wethersfield, CT 06109

**The Cook's Garden**
Box 65
Londonderry, VT 05148

**Dominion Seed House**
Georgetown, Ontario
L7G 4A2 Canada

**Farmer Seed & Nursery Company**
1706 Morrissey Drive
Bloomington, IL 61704

**Henry Field's Seed & Nursery Company**
Shenandoah, IA 51602

**Gardeners' Choice Catalog**
10 South Franklin Turnpike
Ramsey, NJ 07446

**Gurney Seed & Nursery Company**
Yankton, SD 57079

**H. G. Hastings Company**
Box 115535
Atlanta, GA 30310-8535

**Ed Hume Seeds, Inc.**
Box 1450
Kent, WA 98035

**Johnny's Selected Seeds**
Foss Hill Road
Albion, ME 04910

**J. W. Jung Seed Company**
335 South High Street
Randolph, WI 53957-0001

**Kilgore Seed Company**
1400 West First Street
Sanford, FL 32771

**D. Landreth Seed Company**
180-188 West Ostend Street
Baltimore, MD 21230

**Orol Ledden & Sons**
Box 7
Sewell, N.J. 08080-0007

**Le Marche Seeds International**
Box 190
Dixon, CA 95620

**Liberty Seed Company**
Box 806
New Philadelphia, OH 44663

**Lockhart Seeds**
Box 1361
Stockton, CA 95201

**McFayden Seeds**
Box 1030
Minot, ND 58702-1030

**Earl May Seed & Nursery**
208 North Elm Street
Shenandoah, IA 51603

**Mellinger's, Inc.**
2310 West South Range Road
North Lima, OH 44452-9731

**Meyer Seed Company**
600 South Caroline Street
Baltimore, MD 21231

**Harris Moran Seed Company**
3670 Buffalo Road
Rochester, NY 14624

**Nichols Garden Nursery**
1190 North Pacific Highway
Albany, OR 97321

**L. L. Olds Seed Company**
Box 7790
Madison, WI 53707-7790

**George W. Park Seed Company, Inc.**
Greenwood, SC 29647

**The Pepper Gal**
10536 119 Avenue North
Largo, FL 34643
*Peppers only*

**W. H. Perron & Company, Ltd.**
515 Labelle Boulevard
Laval, Quebec H7V 2T3
Canada

**Piedmont Plant Company**
Box 424
Albany, GA 31703

**Porter & Son, Seedsmen**
Box 104
Stephenville, TX 76401-0104

**Redwood City Seed Company**
Box 361
Redwood City, CA 94064

**Rocky Mountain Seed Company**
Box 5204
Denver, CO 80217

**Seeds Blum**
Idaho City Stage
Boise, ID 83706

**Shepherd's Garden Seeds**
6116 Highway 9
Felton, CA 95018

**R. H. Shumway, Seedsman**
Box 1
Graniteville, SC 29829

**Southern Seeds**
Box 2091
Melbourne, FL 32902
*Hot weather varieties*

**Stokes Seeds**
Box 548
Buffalo, NY 14240
*Cold weather varieties*

**Territorial Seed Company**
Box 27
Loranne, OR 97451

**Tomato Growers Supply Company**
Box 2237
Fort Meyers, FL 33902
*Tomatoes and peppers*

**Tsang and Ma International**
Box 5644
Redwood City, CA 94063
*Asian vegetables*

**T & T Seeds, Ltd.**
Box 1710
Winnipeg, Manitoba
R3C 3P6 Canada

**Otis S. Twilley Seed Company, Inc.**
Box 65
Trevose, PA 19047

**Vermont Bean Seed Company**
11 Garden Lane
Bomoseen, VT 05732
*Cold weather varieties*

**Vesey's Seeds, Ltd.**
York, Prince Edward
Island C0A 1P0
Canada

**Wetsel Seed Company, Inc.**
Box 791
Harrisonburg, VA 22801

# INDEX

*Note: Page numbers in boldface type indicate principal references; page numbers in italic type indicate references to illustrations.*


**A**

Acorn squash. *See* Squashes, winter squashes
All-America Selections, 49
*Allium ampeloprasum, Porrum* Group. *See* Onions, leeks
*Allium cepa*
  *Aggregatum* Group. *See* Onions, shallots
  *Cepa* Group. *See* Onions, bulbing onions; Onions, green onions
  *Proliferum* Group. *See* Onions, Egyptian onions
*Allium fistulosum. See* Onions, green onions
*Allium sativum. See* Onions, garlic
*Allium schoenoprasum. See* Onions, chives
*Allium tuberosum. See* Onions, chives
Aluminum foil mulch, 17
*Amaranthus tricolor. See* Spinach
Animals, controlling, 39
Anise. *See* Fennel
Aphids, 33, 37
*Apium graveolens* var. *dulce. See* Celeriac; Celery
*Arachis hypogaea. See* Peanuts
*Armoracia rusticana. See* Horseradish
Artichokes, 26, 63, **70,** *70. See also* Jerusalem artichokes
Arugula, 63, **70,** *71*
Asparagus, **71,** *71*
  pests of, 37
  planting chart for, 63
  storing, 66, 67
  temperature for, 26
  watering, 19 (chart)
Asparagus beans. *See* Beans, yard-long beans
Asparagus beetles, 37
Asparagus lettuce. *See* Celtuce
*Asparagus officinalis. See* Asparagus

**B**

*Bacillus thuringiensis* (Bt), 34
Balsam gourds. *See under* Gourds
*Barbarea verna,* **88**
Bark, ground, 13, 17
*Basella alba. See* Spinach
Beans, **71–76**
  bush beans, 48, 72, *72,* 73, 74-75
  common mistakes when growing, 72
  container-grown, 48
  dry beans, **72**
  fava beans, 63, **72–73,** *73*
  garbanzo beans, 63, 72, **73,** *73*
  heirloom beans, 73
  horticultural beans, 72, **73**
  and intercropping, 44
  lima beans, 19 (chart), 72, **73–74,** *74*
    storing, 66, 67
  pests of, 34, 35
  planting, 72
  planting chart for, 63
  pole beans, 45, 72, 73, 74-75
  scarlet runner beans, 63, **74,** *74*
  snap beans, 72, **74–75,** *75*
    bush, *72,* **74–75**
    container-grown, 48
    disease resistance of, 37
    purple, 74, *75*
    storing, 66, 67
    and water stress, 20
    yellow (wax), 67, 72, 74, 75
  soybeans, 63, 72, **75,** *75*
  sprouting, 75-76
  staking, 45
  storing, 66, 67
  temperature for, 26-27
  wax beans, 67, 72, 74, 75
  yard-long beans, 63, 72, **76,** *76*
Beetles, 33, 37
Beets, **117–18,** *117, 118*
  container-grown, 48
  planting chart for, 63
  storing, 66, 67
  temperature for, 26, 117
  and water stress, 20
Belgian endive. *See* Chicory
Belle Isle cress, **88**
Bell peppers. *See under* Peppers
*Benincasa hispida. See* Gourds, Chinese preserving melons
*Beta vulgaris*
  *Cicla* Group. *See* Chard
  *Crassa* Group. *See* Beets
Birds, controlling, 39, 87
Black-eyed peas. *See* Peas, cowpeas
Black plastic film mulch, 17, 28, 32, 38
Bok choy. *See* Cabbage, pak choi
Bolting, 25-26, 76-77
Bottle gourds. *See under* Gourds
*Brassica campestris. See* Mustard greens
*Brassica juncea. See* Mustard greens
*Brassica oleracea*
  *Acephala* Group. *See* Collards; Kale
  *Botrytis* Group. *See* Broccoli; Cauliflower
  *Capitata* Group. *See* Cabbage
  *Gemmifera* Group. *See* Brussels sprouts
  *Gongylodes* Group. *See* Kohlrabi
*Brassica rapa*
  *Chinensis* Group. *See* Cabbage, pak choi
  *Napograssica* Group. *See* Rutabagas
  *Pekinensis* Group. *See* Cabbage, Chinese cabbage
  *Rapifera* Group. *See* Turnips
Broad beans. *See* Beans, fava beans
Broccoli, **77,** *77*
  bolting of, 76-77
  container-grown, 48
  fertilizing, 76
  pests of, 33, 35
  planting, 76-77
  planting chart for, 63
  storing, 66, 67
  temperature for, 26, 76, 77
  watering, 19 (chart)
Brussels sprouts, **77–78,** *78*
  bolting of, 76-77
  container-grown, 48
  fertilizing, 76
  pests of, 33, 35
  planting, 76-77
  planting chart for, 63
  storing, 66
Bt (*Bacillus thuringiensis*), 34
Bugs, 33. *See also* Insects
Burlap
  as mulch, 17
  as shade source, 60
  as wind protection, 59
Bush beans. *See under* Beans
Bush cucumbers. *See under* Cucumbers
Bush squashes. *See under* Squashes
Butter beans. *See* Beans, lima beans
Butterhead lettuce. *See* Lettuce, types of
Butternut squash. *See* Squashes, winter squashes

**C**

Cabbage, *76, 77,* **78–79,** *79*
  bolting of, 76-77
  Chinese cabbage, 25, 66, 67, **79–80,** *79*
  container-grown, 48
Cabbage (*continued*)
  disease resistance of, 37
  fertilizing, 23, 76, 78
  green cabbage, 78
  how to use, 79
  ornamental flowering, *23*
  pak choi, **81**
  pests of, 33, 34, 35
  planting, 76-77, 78
  planting chart for, 63
  problems with, 78
  red cabbage, *76,* 78
  savoy cabbage, 78, 79
  storing, 66, 67
  temperature for, 26, 76, 78
  watering, 19 (chart)
Cabbage loopers, 34
Cabbage root maggots, 35
Calendar, for typical garden, 43
Cantaloupes. *See under* Melons
Cape gooseberry. *See* Tomatoes, husk tomatoes
*Capsicum annuum. See* Peppers
Cardoon, 63, **81–82,** *81*
Carrots, **118–19,** *118, 119*
  container-grown, 48
  and intercropping, 44
  planting chart for, 63
  and stinkbugs, 33
  storing, 66, 67
  temperature for, 26
  watering, 19 (chart)
Casabas. *See* Melons, winter melons
Caterpillars, 33-34
Cauliflower, *78,* **79,** *79*
  container-grown, 48
  fertilizing, 76
  pests of, 33
  planting, 76-77, 79
  planting chart for, 63
  storing, 66, 67
  temperature for, 26, 76, 79
  watering, 19 (chart)
Celeriac, 63, 66, **82–83,** *82*
Celery, 26, 63, **82–83,** *82*
  storing, 66, 67
Celery cabbage. *See* Cabbage, Chinese cabbage
Celtuce, 63, **83,** *83*
Chard, 48, 63, 67, **83–84,** *83*
Chayote, 63, **84,** *84,* 128 (chart)
Cheesecloth, as shade source, 60
Cherry tomatoes. *See* Tomatoes, husk tomatoes
Chick-peas. *See* Beans, garbanzo beans
Chicons, 85
Chicory, 63, **84–85,** *84, 85. See also* Radicchio
Chile peppers. *See under* Peppers

Chinese cabbage. *See under* Cabbage
Chinese chives. *See* Onions, chives
Chives. *See under* Onions
Chop suey greens. *See* Shungiku
*Chrysanthemum coronarium. See* Shungiku
*Cicer arietinum. See* Beans, garbanzo beans
*Cichorium endiva. See* Endive; Escarole
*Cichorium intybus. See* Chicory; Radicchio
*Citrullus lanatus. See* Melons, watermelons
Climate, **24–28**. *See also* Temperature
Cloches, 29
Clubroot disease, 76
Cold frames, 30, *31*
Cole vegetables. *See specific vegetables*
Collards, 63, 66, 76, **80,** *80*
Colorado potato beetles, 37
Community vegetable gardens, 47
Companion planting, 37
Compost, 13, **15–16**, 17
Container-grown vegetables, 17–18, 20, 24, **47–48**
Cooking vegetables. *See specific vegetables*
Corn, *27,* **85–88**, *85, 86, 87*
  bicolor corn, 88
  fertilizing, 86
  harvesting, 86
  history of, 85
  how to use, 88
  pests of, 34, 87
  planting, 45, 85–86
  planting chart for, 63
  popcorn, 67, 85, 86, 88
  storing, 66, 67
  temperature for, 26, 85–86
  tips on, 86–87
  varieties of, 87–88
  watering, 19 (chart), 86
  white corn, 87
  yellow corn, 87
Corn earworms, 34, 87
Cos lettuce. *See* Lettuce, types of
Cowpeas. *See under* Peas
Crenshaw melons. *See* Melons, winter melons
Cress, 63, **88–89,** *88*
Crisphead lettuce. *See* Lettuce, types of
Crop rotation, 37
Cucumber beetles, 33
Cucumbers, *88,* **89–92,** *89, 90, 91,* 128 (chart)
  bitterness in, 90
  bush cucumbers, 48, 90, 92
  container-grown, 48, 93
  disease resistance of, 37, 91
Cucumbers (*continued*)
  gynoecious, 90-91
  harvesting, 90, 93
  how to use, 92
  pests of, 33, 35, 93
  pickling cucumbers, 90, 91-92, *91*
  planting, *53,* 89-90, 93
  planting chart for, 63
  slicing cucumbers, 90, 91
  spine color of, 90
  storing, 66, 67
  temperature for, 26, 89, 93
  varieties of, 90-92
  watering, 19 (chart), 20
*Cucumis melo*
  *Inodorus* Group. *See* Melons, winter melons
  *Reticulatus* Group. *See* Melons, cantaloupes; Melons, muskmelons
*Cucumis sativus. See* Cucumbers
*Cucurbitaceae,* **124–30.** *See also* Gourds; Pumpkins; Squashes
Curly Cress. *See* Garden cress
Cutworms, 34
*Cynara cardunculus. See* Cardoon
*Cynara scolymus. See* Artichokes

**D**
Daikon radishes. *See under* Radishes
Damping-off disease, 56
Dandelions, 64, **92**
*Daucus carota* var. *sativus. See* Carrots
DCPA, 38
Deer, controlling, 39
Depth, to plant seed, 56, 59, 63-65 (chart)
*Diabrotica,* 33
Diseases, controlling, 29, 32, **37**. *See also specific diseases*
Dishrag gourds. *See* Gourds, luffa gourds
Distance, between plants/rows, 63-65 (chart)
Dogs, controlling, 39
Dolomitic limestone, 16
Double cropping, 43–44
Double-digging, 15
Drip irrigation, **20–21**, 28
Dry beans. *See under* Beans
Drying
  gourds, 127
  peppers, 113
  vegetables, 66

**E**
Early winter cress, **88**
Eggplant, **92–94,** *92, 93*
  container-grown, 48
  planting chart for, 63
Eggplant (*continued*)
  starting seeds for, 61
  storing, 66, 67
  temperature for, 26-27
  varieties of, 93
  watering, 19 (chart)
Egyptian onions. *See under* Onions
Elephant garlic. *See* Onions, garlic
Endive, **94,** *94*
  French or Belgian. *See* Chicory
  planting chart for, 63, 64
  storing, 66, 67
English peas. *See* Peas, garden peas
*Eruca vesicaria sativa. See* Arugula
Escarole, 66, **94.** *See also* Endive

**F**
Fava beans. *See under* Beans
Fennel, 64, **94–95,** *94*
Fertilizers, 8-9, **22–24**
Fiber pots, 55
Film mulches, 17, 28, 32, 38
Film row covers, 28, 29, 32
Flageolets. *See* Beans, horticultural beans
Floating row covers, 28, *29*
Florence fennel. *See* Fennel
*Foeniculum vulgare* var. *azoricum. See* Fennel
Freezing, vegetables, 66–67
*Freezing & Drying* (Ortho), 66
French endive. *See* Chicory
Frost, *26,* 28–30, 59
Fungus diseases, 37
Furrow irrigation, 20

**G**
Garbanzo beans. *See under* Beans
Garden beans. *See* Beans, snap beans
Garden cress, 63, **88–89**
Garden huckleberry, 64, **95**
Garden peas. *See under* Peas
Gardens. *See* Vegetable gardens
Garland chrysanthemum. *See* Shungiku
Garlic. *See under* Onions
Garlic chives. *See* Onions, chives
Germination, 52–54, 55–56, 59–60, 61, 63–65 (chart)
Globe artichokes. *See* Artichokes
*Glycine max. See* Beans, soybeans
Glyphosate, 38
Gooseberry tomatoes. *See* Tomatoes, husk tomatoes
Gophers, controlling, 39
Gourds, 64, 124-25, **127–30,** *127–30*
Gourds (*continued*)
  balsam gourds, 128 (chart), *128*
  bottle gourds, 127, 128 (chart), 129, *130*
  Chinese preserving melons, 128 (chart)
  crossing squashes with, 129–30
  drying, 127
  luffa gourds, **127–29,** *129*
  planting, 124–25, 127
  snake gourds, 128 (chart), **129**
Grams. *See* Beans, garbanzo beans
Grass clippings, as compost material, 15, 16
Green beans. *See* Beans, snap beans
Greenhouses
  mini, 28, *28,* 60
  trench, 61
Green onions. *See under* Onions
Green peas. *See* Peas, garden peas
Ground-cherry. *See* Tomatoes, husk tomatoes
Growing season
  extending, 9, **28–30,** *31*
  length of, 9, 24–25
Gumbo. *See* Okra
Gypsum, 13

**H**
Hardening off, transplants, 58
Hardware cloth, 59
Harvesting, 42, **65–67**
Heat, how gardens gain and lose, 27, 28, 29
*Helianthus annuus. See* Sunflowers
*Helianthus tuberosus. See* Jerusalem artichokes
Herbicides, 38
Honeydew melons. *See* Melons, winter melons
Horse beans. *See* Beans, fava beans
Horseradish, 64, 66, **95–96,** *95, 96*
Horticultural beans. *See* Beans, horticultural beans
Hotbeds, 30
Hot caps, 59
Hubbard squash. *See* Squashes, winter squashes
Huckleberry. *See* Garden huckleberry
Husk tomatoes. *See under* Tomatoes
Hybrid vegetables, 49

**I, J**
Iceberg lettuce. *See* Lettuce, types of
Insecticides, 32
Insects, controlling, 17, 29, **32–37**. *See also specific insects*
Intensive gardening, 46
Intercropping, 43–44
*Ipomoea batatas*. *See* Sweet potatoes
Irrigation. *See* Watering
Japanese peppers. *See under* Peppers
Jerusalem artichokes, 64, *65*, **96–97**, *96*
Jicama, 64, **97**, *97*

**K**
Kale, 26, 64, 76, **80–81**, *80*
 storing, 66, 67
Kohlrabi, *9*, **81**, *81*
 bolting of, 76–77
 container-grown, 48
 fertilizing, 76
 planting, 76–77, 81
 planting chart for, 64
 storing, 66, 67
 temperature for, 26, 76

**L**
Labeling plants, 49
*Lactuca sativa*. *See* Celtuce; Lettuce
*Lagenaria siceraria*. *See* Gourds
Laying out, vegetable gardens, 45–46
Leafhoppers, 34–35
Leaf lettuce. *See* Lettuce, types of
Leaf miners, 35
Leaves
 as compost material, 15
 as mulch, 17, 32
Leeks. *See under* Onions
Length
 of day, 25–26
 of growing season, 9, 24–25
*Lepidium sativum*. *See* Garden cress
Lettuce, *24*, **97–99**, *97, 98, 99*
 container-grown, *47*, 48
 extending growing season of, 30
 harvesting, 98
 how to use, 99
 pests of, 33, 34, *39*
 planting, 97–98
 planting chart for, 64
 storing, 66, 67
 temperature for, 26, 97–98
 types of, 98–99
 watering, 19 (chart), 20
Light
 amount required, 11–12
 and day length, 25–26
 and fertilizing, 23
 and seedlings, 54, 59, 60
Light (*continued*)
 and trapping heat, 27, 28, 29
Lima beans. *See under* Beans
Lime, hydrated, 16
Limestone, crushed, 16
Luffa gourds. *See under* Gourds
*Lycopersicon lycopersicum*. *See* Tomatoes

**M**
Manure, 13, 17
Marigolds, as pest repellent, 37
Maturity, days to, 63–65 (chart)
Melons, **99–102**, *99, 100, 101*, 128 (chart)
 cantaloupes, 66, 99, **100**, *100*
 muskmelons, 99, **100**
  disease resistance of, 37, 100
  planting chart for, 64
  space-saving, 49
  storing, 67
  temperature for, 26
  and water stress, 20
 pests of, 35
 planting, *53*, 99
 storing, 66, 67
 tips on, 100
 watering, 19 (chart)
 watermelons, 26, 65, **100–1**, *100*
  storing, 66, 67
 winter melons, 99, **101–2**, *101*
Mexican bean beetles, 37
Mexican potatoes. *See* Jicama
Mice, controlling, 39
Microclimate, 24, 27
Mirliton. *See* Chayote
Mites, 35, 37
Moisture. *See* Watering
Moles, controlling, 39
*Momordica*. *See* Gourds, balsam gourds
Mulches, *16*, **17**, 27, 28, 32, 38
Muskmelons. *See under* Melons
Mustard greens, 64, 67, **102**, *102*

**N, O**
*Nasturtium officinale*. *See* Watercress
Nasturtiums, 37, 64, **102–3**, *102*
National Garden Bureau, 44
Nematodes, 37
Newspapers, shredded, 32
Okra, **103**, *103*
 planting chart for, 64
 and stinkbugs, 33
 storing, 66
Okra (*continued*)
 temperature for, 26, 27, 103
Onion maggots, 35
Onions, *69*, **103–7**, *103–7*
 bulbing onions, **103–5**, *103*
 chives, 30, 63, **105**, *105*
 and day length, 25, 104
 Egyptian onions, **105**
 fertilizing, 23, 104
 garlic, *69, 103, 104*, **105–6**
  as aphid repellent, 37
  planting chart for, 64
  storing, 66
  temperature for, 26
 green onions (scallions), *103*, 104, **106**, *106*
 harvesting, 104
 history of, 103
 and intercropping, 44
 leeks, 26, 64, 66, *103*, **106–7**, *106*
 pests of, 35
 planting, 104
 planting chart for, 64
 shallots, 26, 66, **107**, *107*
 storing, 66, 67
 temperature for, 26
 watering, 19 (chart)
Organic fertilizers, 22, 23. *See also* Compost; Manure
*The Ortho Problem Solver* (Ortho), 32
Oyster plant. *See* Salsify

**P**
*Pachyrrhizus erosus*. *See* Jicama
Pak choi. *See under* Cabbage
Paper mulch, 17
Parsnips, **119–20**, *119*
 container-grown, 48
 planting chart for, 64
 storing, 66, 67
 temperature for, 26
*Pastinaca sativa*. *See* Parsnips
Peanuts, 64, **107–8**, *107*
Peas, **108–10**, *108, 109*
 cowpeas (black-eyed peas), 63, **108–9**, *108, 109*
 garden peas (English peas), **108**, *108*
 pests of, 35
 planting chart for, 63, 64
 storing, 66, 67
 sugar peas (snow peas), **109**
 sugar snap peas, *8*, **109–10**, *109*
Peat moss, 13, 17, 18, 56
Peat pellets, 55
Peat pots, 55, 59
Peppergrass. *See* Garden cress
Peppers, *69*, **110–13**, *110–13*
 bell peppers, **110–11**, *110, 111*
 chile peppers, 48, 66, **112–13**, *112, 113*
 container-grown, 48
Peppers (*continued*)
 drying, 113
 harvesting, 110
 Japanese peppers, **113**
 pests of, 33, 34, 35
 pimientos, **112**
 planting, 110
 planting chart for, 64
 starting seed for, 61, *62*
 storing, 66, 67
 temperature for, 26, 110
Perlite, 17, 18, 56
Persian melons. *See* Melons, winter melons
Pesticides, 32
Pests
 *See also specific pests*
 animals and birds, 39
 insects, 17, 29, **32–37**
pH, of soil, 16
*Phaseolus coccineus*. *See* Beans, scarlet runner beans
*Phaseolus lunatus*. *See* Beans, lima beans
*Phaseolus vulgaris*. *See* Beans, horticultural beans; Beans, snap beans
*Physalis ixocarpa*. *See* Tomatillos
*Physalis peruviana*. *See* Tomatoes, husk tomatoes
Pimientos. *See under* Peppers
*Pisum sativum*. *See* Peas
Planning vegetable gardens, 7–8, **41–49**
 deciding what to plant, 44
 and harvest period length, 42
 and intercropping, 43–44
 necessity for, 41–42
 sample plan, 44–45
 and succession plantings, 43
Planting chart, 62–65
Planting mixes, 17–18
Planting vegetables, 51–65
 and laying out garden, 45–46
 planting chart for, 62–65
 seeds, *51*, 52–54
  starting indoors, 54–59
  starting outdoors, 59–61
  transplants vs., 51
 transplants
  hardening off, 58
  protecting, 59
  purchasing, 62
  seeds vs., 51
  setting out, *57*, 58–59, *58*
Plastic film mulches, 17, 28, 32, 38
Plastic jugs, to protect plants, 29, *31*
Plastic row covers, 28, 29, 32
Plugs, 55

Poha. *See* Tomatoes, husk tomatoes
Pole beans. *See under* Beans
Popcorn. *See under* Corn
Potatoes, **113–16,** *114, 115, 116*
  *See also* Sweet potatoes
  common mistakes when growing, 114-15
  container-grown, *18*
  harvesting, 114
  how to use, 116
  new potatoes, 114
  pests of, 34
  planting, 114, 115
  planting chart for, 64
  storing, 66, 67
  temperature for, 26, 113
  varieties of, 115-16
Poultry, controlling, 39
Protecting vegetables
  from animals and birds, 39, 87
  from elements, 9, 28-30, *31,* 32, 59
Pumice, 17
Pumpkins, **124–25**, 126, 127, *127,* 128 (chart)
  how to use, 127
  planting, *53,* 124-25
  planting chart for, 64
  space-saving, 49
  and stinkbugs, 33
  storing, 66, 67
  temperature for, 26, 124
  varieties of, 126

**R**
Rabbits, controlling, 39
Radicchio, 64, **116,** *116*
Radishes, *69,* **120–21,** *120*
  container-grown, 48
  daikon radishes, 120-21
  and intercropping, 44
  as pest repellent, 37
  pests of, 35
  planting chart for, 64
  storing, 66, 67
  and water stress, 20
Radish maggots, 35
Raised beds, 17, 18, **46**
*Raphanus sativus. See* Radishes
Record keeping, 49
*Rheum rhabarbarum. See* Rhubarb
Rhubarb, 26, 64, **116–17,** *117*
  storing, 66, 67
Rocket. *See* Arugula
Rodents, controlling, 39
Romain lettuce. *See* Lettuce, types of
Root vegetables. *See specific vegetables*
Rotary tilling, 13-15
Rotating crops, 37
Row covers, 28-29
*Rumex acetosa. See* Sorrel
Rutabagas, 26, 64, **121–22**
  storing, 66, 67

**S**
Salad gardens, vegetables for, 44
Salsify, 48, 64, **121,** *121*
Salt hay, 32
Sand, coarse, 17, 18
Savoy cabbage. *See under* Cabbage
Sawdust, 13, 15
Scallions. *See* Onions, green onions
Scarlet runner beans. *See under* Beans
*Scorzonera. See* Salsify
*Sechium edule. See* Chayote
Seed flats, 55
Seedlings
  caring for, 54, 58
  hardening off, 58
  thinning, 60-61
  transplanting, *57,* 58–59, *58*
Seeds
  germination of, 52-54, 55-56, 59-60, 61, 63-65 (chart)
  growing plants from, *51,* 52-54
  planting chart for, 62-65
  sources of, 139
  starting indoors, 54-59
  starting outdoors, 59-61
  storing, 61
  testing germination of, 61
  transplants vs., 51
Serpent cucumbers. *See* Gourds, snake gourds
Shade, 11-12, 59, 60. *See also* Light
Shallots. *See under* Onions
Shell beans. *See* Beans, dry beans
Shredders, 15
Shungiku, 64, **122**
Single-row planting system, 46
Slope of land, climate and, 27-28
Slugs, 35
Small-space vegetable gardens, 47-49
Snails, 35
Snap beans. *See under* Beans
Snow peas. *See* Peas, sugar peas
Soil, **12–18**
  air content of, 12-13
  clay, 12-13, 15, 19
  and compost, 15-16
  in containers, 17-18
  and germination, 52, 63-65 (chart)
  improving, 13, 15
  leveling, *14*
  and mulches, 17
  optimal, 13, 15
  pH of, 16
  in raised beds, 17, 18
  rotary tilling, 13-15
Soil (*continued*)
  sandy, 12, 13, 19
  synthetic, 17-18
  temperature of, 52, 62, 63-65 (chart)
  testing, 16-17
  types of, 12-13
  warming, 27, 28, 29
  water content of, 12-13, 15
  working, 13-15
Soil blocks, 55-56, 59
*Solanum melanocerasum. See* Garden huckleberry
*Solanum melongena. See* Eggplant
*Solanum tuberosum. See* Potatoes
Sorrel, 64, **122–23**
Sour grass. *See* Sorrel
Southern peas. *See* Peas, cowpeas
Sowing seeds
  indoors, 56, *57*
  outdoors, 59-60, 61
Soybeans. *See under* Beans
Spacing, of plants/rows, 63-65 (chart)
Spaghetti squash. *See* Squashes, winter squashes
Sphagnum moss, 56
Spinach, *22,* **123–24,** *123, 124*
  bolting of, 25, 123
  and day length, 25-26
  disease resistance of, 37
  planting chart for, 65
  storing, 66, 67
  temperature for, 26
*Spinacia oleracea. See* Spinach
Sprinklers, 21, *37*
Squash bugs, 33, 37
Squashes, **124–27,** *124-26*
  bush squashes, 48-49
  crossing gourds with, 129-30
  harvesting, 125
  how to use, 126-27
  pests of, 33, 34, 35, 37
  planting, *53,* 124-25
  planting chart for, 65
  storing, 66, 67
  summer squashes, *27, 69,* **124–27,** *125,* 128 (chart)
  temperature for, 26
  winter squashes, **124–25,** *124,* **126,** *126,* 127, 128 (chart), *129*
Staking, 30-32
Stinkbugs, 33
Storing
  seeds, 61
  vegetables, 65-67
Straw, *16,* 17, 32
Strawberry tomatoes. *See* Tomatoes, husk tomatoes
String beans. *See* Beans, snap beans
Succession plantings, 43
Sugar peas. *See under* Peas
Sugar snap peas. *See under* Peas
Sulfur, 16
Summer squashes. *See under* Squashes
Sunchokes. *See* Jerusalem artichokes
Sunflowers, 64, **130,** *130*
Sunlight. *See* Light
Supports, 30-32
Sweet anise. *See* Fennel
Sweet corn. *See* Corn
Sweet peppers. *See* Peppers
Sweet potatoes, *61,* **130–32,** *131, 132*
  planting chart for, 65
  storing, 66
  temperature for, 26
Swiss chard. *See* Chard

**T**
Tampala. *See* Spinach
*Taraxacum officinale. See* Dandelions
Temperature
  *See also* Climate; Heat
  for growing vegetables, 26-27
  and seed germination, 52
*Tetragonia tetragonioides. See* Spinach
Thinning, 60-61
Tilling, 13-15
Tomatillos, 65, **132,** *132*
Tomatoes, *49, 69,* **132–38,** *133-35, 138*
  bush tomatoes, 133, 134, 135
  caging, 32, 133, 134
  common mistakes when growing, 135
  container-grown, 48, 49, 134
  determinate, 133, 135
  disease resistance of, 37, 135, 136-37 (chart)
  fertilizing, 133
  history of, 132
  how to use, 135-38
  husk tomatoes (ground-cherry), 64, **96,** *135*
  indeterminate, 133, 135
  intermediate short internode (ISI), 133, 135
  pests of, 34, 35
  planting, 132-33
  planting chart for, 64
  problems with, 134-35
  protecting, 32, 59, 134
  pruning, 134
  staking, 32, 45, 134
  starting seed for, 61, 133
  storing, 66, 67, 138
  temperature for, 26-27, 132
  varieties of, 135, 136-37 (chart)
  watering, 19 (chart), 133

Tomato fruitworms, 34
Tomato hornworms, 34
Tools, basic gardening, 12
*Tragopogon porrifolius. See* Salsify
Transplants
- hardening off, 58
- protecting, 59
- purchasing, 62
- seeds vs., 51
- setting out, *57,* 58-59, *58*

Trench greenhouses, 61
*Trichosanthes anguina. See* Gourds, snake gourds
*Tropaeolum majus. See* Nasturtiums
Turban squash. *See* Squashes, winter squashes
Turnips, *118,* **121-22,** *122*
- container-grown, 48
- planting chart for, 65
- storing, 66, 67
- temperature for, 26, 121
- and water stress, 20

**U, V**

Upland cress, **88**
Vegetable gardens, *6*
- *See also* Vegetables
- avoiding failures in, 6-7
- calendar for, 43
- choosing vegetables for, 49
- and climate, 24-28
- community gardens, 47
- companion planting in, 37
- disease control in, 29, 32, **37**

Vegetable gardens (*continued*)
- fertilizing, 8-9, **22-24**
- following directions vs. instinct in, 8-9
- harvesting, 42, 65-67
- insect control in, 17, 29, **32-37**
- labeling plants in, 49
- laying out, 45-46
- planning, 7-8, **41-49**
- planting, 51-65
- record keeping for, 49
- small-space, 47-49
- soil in, 12-18
- tools required for, 12
- watering, *10,* **18-22**
- weeds in, 37-38, 60

Vegetable pears. *See* Chayote
Vegetables
- *See also specific vegetables*; Vegetable gardens
- container-grown, 17-18, 20, 24, **47-48**
- cool-season, 26-27
- drying, 66
- exotic, 9
- freezing, 66-67
- growing season of,
  - extending, 9, **28-30,** *31*
  - length of, 9, 24-25
- homegrown vs. store-bought, 9
- how to use. *See specific vegetables*

Vegetables (*continued*)
- hybrid, 49
- light requirements of, 11-12
- protecting
  - from animals and birds, 39, 87
  - from elements, 9, 28-30, *31,* 32, 59
- reasons for growing, 5-6
- storing, 65-67
- supports for, 30-32
- varieties of, 49
- warm-season, 26-27

Vermiculite, 17, 56
*Vicia faba. See* Beans, fava beans
*Vigna unguiculata. See* Peas, cowpeas
*Vigna unguiculata sesquipedalis. See* Beans, yard-long beans

**W, X, Y, Z**

Watercress, *88,* **89**
Watering, *10,* **18-22**
- container-grown vegetables, 20
- critical period for, 19 (chart)
- and germination, 52, 53, 60
- methods of, 20-22
  - drip irrigation, **20-21,** 28
  - furrow irrigation, 20
  - by hand, 21-22
  - sprinklers, 21
- overwatering, 18

Watering (*continued*)
- procedure for, 20
- quantity of, 20
- and row covers, 28
- and soil type, 12-13, 15, 19
- timing of, 19 (chart), 20
- underwatering (water stress), 19-20

Watermelons. *See under* Melons
Wax beans. *See under* Beans
Weeds, **37-38,** 60
Whiteflies, 35-36
Wide-row planting system, 46
Wind protection, 28, 59
Windsor beans. *See* Beans, fava beans
Winter cherry. *See* Tomatoes, husk tomatoes
Winter melons. *See under* Melons
Winter protection, 32
Winter squashes. *See under* Squashes
Witloof. *See* Chicory
Worms, 33-34
Yam beans. *See* Jicama
Yams, 130. *See also* Sweet potatoes
Yard-long beans. *See under* Beans
Yellows disease, 78
*Zea mays* var. *rugosa. See* Corn
Zucchini. *See* Squashes, summer squashes


## *U.S. Measure and Metric Measure Conversion Chart*

| | | Formulas for Exact Measures | | | Rounded Measures for Quick Reference | | |
|---|---|---|---|---|---|---|---|
| | Symbol | When you know: | Multiply by: | To find: | | | |
| Mass (Weight) | oz | ounces | 28.35 | grams | 1 oz | | = 30 g |
| | lb | pounds | 0.45 | kilograms | 4 oz | | = 115 g |
| | g | grams | 0.035 | ounces | 8 oz | | = 225 g |
| | kg | kilograms | 2.2 | pounds | 16 oz | = 1 lb | = 450 g |
| | | | | | 32 oz | = 2 lb | = 900 g |
| | | | | | 36 oz | = 2¼ lb | = 1000g (1 kg) |
| Volume | pt | pints | 0.47 | liters | 1 c | = 8 oz | = 250 ml |
| | qt | quarts | 0.95 | liters | 2 c (1 pt) | = 16 oz | = 500 ml |
| | gal | gallons | 3.785 | liters | 4 c (1 qt) | = 32 oz | = 1 liter |
| | ml | milliliters | 0.034 | fluid ounces | 4 qt (1 gal) | = 128 oz | = 3¾ liter |
| Length | in. | inches | 2.54 | centimeters | ⅜ in. | = 1 cm | |
| | ft | feet | 30.48 | centimeters | 1 in. | = 2.5 cm | |
| | yd | yards | 0.9144 | meters | 2 in. | = 5 cm | |
| | mi | miles | 1.609 | kilometers | 2½ in. | = 6.5 cm | |
| | km | kilometers | 0.621 | miles | 12 in. (1 ft) | = 30 cm | |
| | m | meters | 1.094 | yards | 1 yd | = 90 cm | |
| | cm | centimeters | 0.39 | inches | 100 ft | = 30 m | |
| | | | | | 1 mi | = 1.6 km | |
| Temperature | °F | Fahrenheit | ⅝ (after subtracting 32) | Celsius | 32°F | = 0°C | |
| | °C | Celsius | ⁹⁄₅ (then add 32) | Fahrenheit | 212°F | = 100°C | |
| Area | in.$^2$ | square inches | 6.452 | square centimeters | 1 in.$^2$ | = 6.5 cm$^2$ | |
| | ft$^2$ | square feet | 929.0 | square centimeters | 1 ft$^2$ | = 930 cm$^2$ | |
| | yd$^2$ | square yards | 8361.0 | square centimeters | 1 yd$^2$ | = 8360 cm$^2$ | |
| | a. | acres | 0.4047 | hectares | 1 a. | = 4050 m$^2$ | |